ROUTLEDGE LIBRARY EDITIONS:
MODERN WORLD ECONOMY

Volume 7

THE INTERNATIONAL STEEL INDUSTRY

First published in 1989 by Routledge

This edition first published in 2017
by Routledge
2 Park Square, Milton Park, Abingdon, Oxon OX14 4RN

and by Routledge
605 Third Avenue, New York, NY 10017

Routledge is an imprint of the Taylor & Francis Group, an informa business

British Library Cataloguing in Publication Data
A catalogue record for this book is available from the British Library

ISBN: 978-1-138-63020-8 (Set)
ISBN: 978-1-315-20964-7 (Set) (ebk)
ISBN: 978-1-138-63068-0 (Volume 7) (hbk)
ISBN: 978-1-315-20926-5 (Volume 7) (ebk)

Publisher's Note
The publisher has gone to great lengths to ensure the quality of this reprint but points out that some imperfections in the original copies may be apparent.

Disclaimer
The publisher has made every effort to trace copyright holders and would welcome correspondence from those they have been unable to trace.

THE INTERNATIONAL STEEL INDUSTRY

Restructuring, state policies and localities

RAY HUDSON AND DAVID SADLER

Routledge
Taylor & Francis Group
LONDON AND NEW YORK

The International Steel Industry

Restructuring, state policies and localities

Ray Hudson and David Sadler

ROUTLEDGE
London and New York

First published 1989
by Routledge
11 New Fetter Lane, London EC4P 4EE
29 West 35th Street, New York NY 10001

Typeset by Pat and Anne Murphy, Highcliffe-on-Sea, Dorset
Printed in Great Britain by
Billing & Sons Ltd, Worcester

British Library Cataloguing in Publication Data

Hudson, Ray
The international steel industry:
restructuring, state policies and
localities.
1. Steel industries
I. Title II. Sadler, David
338.4'7669142

ISBN 0-415-02186-3

Library of Congress Cataloging in Publication Data

Hudson, Raymond
The international steel industry: restructuring, state policies, and localities / Ray Hudson and David Sadler.
p. cm.
Bibliography: p.
Includes index.
ISBN 0-415-02186-3
1. Steel industry and trade. 2. Steel industry and trade — Government policy. I. Sadler, David. II. Title.
HD9510.5.H83 1989 88-36530 CIP
338.4'7669142—dc19

Contents

List of figures

List of tables

Preface

Nothing symbolizes the cruelly destructive effects of de-industrialization more than the decaying former steel towns of the USA and north-western Europe. Places such as Consett, Longwy, and Youngstown have become potent symbols of the localized economic and social consequences and costs of industrial decline. Towns that were once central to the accumulation process — indeed, owe their very existence to it — and which provided relatively highly paid jobs for male workers, as well as handsome profits for capitalists, are now excluded from the main currents of capital's ebb and flow. They are characterized by depression, mass unemployment, and the poverty that comes with it. Despite a plethora of state policies to encourage or provide alternative sources of waged labour, they have become stagnant backwaters, of only marginal attraction to capital.

Why did this happen? In what sense, if any, was it inevitable or necessary? Such questions are important, not least for the people who live in these stricken communities. To begin to answer them, we felt it necessary to situate the plight of the former steel towns in the old industrial countries within a changing geography of international production and trade. For as some towns have ceased to be centres of production for steel (and profits), other new production bases have sprung up in parts of the Middle East and Third World, in some of the newly industrializing countries. The decline of steel production in particular locations in the old industrialized countries can be understood only in relation to this expansion elsewhere.

But this is not to imply that the collapse of steel production in places such as Consett is in some sense mechanistically predetermined via changes in the international division of labour in steel production. Instead, we would argue that the rise and fall of these steel towns must be understood in terms of their relationship to this changing international setting mediated through the strategies of the various social actors that seek to influence the geography of production, such as private capitals, central and local governments, and trade unions.

Unravelling these complex relationships between processes of international and national industrial restructuring, and economic and social change in particular places, has come to occupy a prominent position in social scientific research in the 1980s. The old neoclassically based explanations of industrial location, exemplified by Weber in relation to the steel industry, have long been overtaken by events. From the late 1970s the interests of human geographers and social theorists converged upon the social production of places and the constitution of societies in space. Industrial location and change came to be seen as an integral part of social processes and class restructuring.

If theoretical approaches were altering radically, so too was the social reality that theories had to address. As capital became increasingly internationalized, the capacity of national states to influence its activities and manage the trajectory of change within their national territories declined. This was particularly apparent in relation to steel: existing centres of production collapsed, with national governments seemingly powerless to prevent this, while new planned complexes were abandoned. Perhaps the most poignant reminder of this is the abandoned site of Gioia Tauro in the Mezzogiorno, intended to be Italy's fifth integrated steel production complex but now marked only by a massive harbour in a desolate area that will probably never be used for the purpose for which it was intended. The consequences for places such as these which never became steel production sites as well as those that were, and came to be rendered redundant in the new geography of steel production, are profound.

It is important to remember also that parallel processes of spatial reorganization and social change have increasingly become evident in industries as diverse as coalmining, cars, and chemicals. Although we focus here upon steel, the issues posed by these processes of restructuring now have a much more general relevance, raising similar theoretical and political issues.

In writing this book, we have had assistance of various sorts from many sources. The Economic and Social Research Council have provided financial support in several ways, initially via a linked research studentship for David Sadler and later via research grants for work in the UK and France that encompassed the steel industry. The Nuffield Foundation have also provided financial support to examine the steel industry in West Germany. We gratefully acknowledge this financial support. Some of the results and conclusions in this book are indirect by-products of research projects which we conducted from 1984 to 1987 for Cleveland County Council and Sheffield City Council. We are grateful to the members and officers who made up the respective steering committees and provided numerous words of advice and encouragement. Many other people have given us advice and assistance over the years, for which we are also extremely grateful; we owe a special word of thanks to Huw Beynon. This will be translated into material form on a future trip to Bremen! The secretarial and technical staff in the Geography Departments at Durham and Lampeter have been of immense help in producing this book and we are pleased to acknowledge their efforts. As always we must end with the usual disclaimer: the faults are ours alone.

Ray Hudson
David Sadler

Durham and Lampeter
July 1988

Chapter one

Steel, the world economy, and locality: some introductory remarks

Introduction

Consett: clouds darken.

(*Newcastle Journal* 5 December 1979)

The fight for Consett: steel men warn of ghost town danger.

(*Newcastle Journal* 12 December 1979)

Steel closure leaves Consett without hope.

(*Financial Times* 13 December 1979)

Consett: an island of unemployment in a sea of unemployment.

(*Sunday Times Colour Supplement* 1980)

Job creation fails to stem unemployment tide.

(*Financial Times* 6 March 1984)

Consett jobless learning skills of life on the dole.

(*Guardian* 25 May 1984)

Consett has seen the future — and it's workless.

(*Sunday Times Colour Supplement* 15 September 1985)

The steel industry has played a prominent and often symbolic role in the twin processes of de-industrialization and re-industrialization which have characterized the period since the late 1970s in many advanced economies. Equally, many of the communities affected by these changes have become the focal point of media attention and political debate. One of these is a small ex-steelworks town in County Durham, north-east England. It is where, in one sense, this book has its origins. For this modest settlement — Consett — epitomizes many shifting and powerful elements of conflict and change within contemporary capitalist society. In part as a consequence, it has had more than its fair share of academic analysis as well as media coverage in recent years. The newspaper headlines quoted above encapsulate such exposure and its changing emphases; there have been many more in a similar vein.

But such images are, of course, necessarily a simplifying reflection of a simplifying story-line. They provide an inevitably brief snapshot of change frozen at one moment in time, yet they also incorporate a range of more subtle pressures. We reprint them here as much for what they keep from the agenda and conceal as for what they reveal. The political origins of such image construction are deep and often under-estimated.

Equally significant, and our principal concern in this book, are the reasons for change in particular places such as Consett. Undoubtedly these must encompass features specific to each locality but they must be understood within a broader perspective that encompasses capitalism as a global system of commodity production and exchange. The demise of steel production in places like Consett has its origins in the international sphere and understanding change in them has to be situated in the context of the growing and changing forms of internationalization of steel production and trade. Equally the ways in which such closures have been contested, and re-industrialization strategies adopted in and for these localities, have been important in legitimizing the pursuit of private profit as the main economic steering mechanism in capitalist society.

Thus, explaining the collapse of steel employment in places such as Consett in the UK, or Longwy in France, or Youngstown in the USA has to start from a global perspective with the restructuring of this key manufacturing sector, mediated in various ways through the policies of national states, and on occasion an embryonic supranational state. It also has to incorporate a more local dimension and address issues such as: exactly how steel closures came about; why campaigns fought in an often bitter attempt to prevent closure almost always resulted in failure; and why political strategies devised nationally and fought for locally to compensate for the loss of steel jobs came to display such an extraordinary degree of similarity in different countries and continents.

Finally, acknowledging that it is impossible to forecast or predict with any precision in capitalist societies, the future of steel and ex-steel towns remains an uncertain one. For globalized manufacturing sectors such as steel, for communities currently dependent upon these industries, and for those which have witnessed the potentially and actually devastating consequences of capital's international mobility — what does the future hold, especially when it seems that increasingly many national states are encouraging this mobility? Rather than attempt to regulate its effects, these states appear to be rewriting the score via national policies which enable capital to dance around the world at an ever-accelerating pace while local communities and governments are powerless to prevent it. What sort of prospect does this hold out for the former steel towns and current steel towns?

These, then, are the issues we seek to cover in this book. In the introductory chapter, we expand upon our aims and objectives and introduce some key issues. How have processes of internationalization set the context for localized restructuring? How have geographies of production evolved — and why? What are the implications of this for old steel towns such as Consett?

The changing international political and economic context of steel production

The early phases of steel production in the original centres of capitalist industrialization in north-west Europe and the USA were closely linked to the growth of other industries within the national territory. Indeed, such intersectoral input-output linkages were often constructed within the same region, in the process helping create that region. Moreover, especially in the earlier phases of growth in the nineteenth century, these physical input-output relationships were paralleled by those of ownership and possession as steel production took place within large oligopolistic conglomerates, tied together via complex webs of financial linkages (such as interlocking shareholdings, ownerships, and directorships), which grew rapidly via mergers and take-overs as part of a process of centralization of capital.[1] Physical input-output linkages extended both backwards and forwards: backwards to raw material supplies (notably coking coals and iron ores) and forwards to industries which consumed steel as an input to their production processes. Initially in the nineteenth century, these forward linkages were to industries such as railway and railway engine manufacture, shipbuilding and related activities and a whole range of mechanical engineering industries. Moreover, steel production remained crucial in the production of the means of destruction. Consequently national governments were extremely anxious to ensure that what was also a strategically significant industry was prominently developed within their national territories. Dependence upon imports of what was in many ways the key raw material for the defence industries was unthinkable for the major industrial and/or colonial powers and would-be powers.

This is not to say that no steel was exported or internationally traded in this early period. In so far as it was, directly and indirectly, it was very much within the context of an old international division of labour. During the 1920s and 1930s the upsurge of protectionist trade measures made national markets all the more important, especially those for new growth industries such as consumer goods and motor vehicles. After 1945, however, things began to change. The post-war period was characterized by a liberalization of international trade within the new politically negotiated trade and monetary framework of Bretton Woods and the General Agreement on Tariffs and Trade. Although this new international setting had little immediate effect, it ushered in the long post-war boom: growing steel demand, boosted initially by the war economy and post-war reconstruction, was sustained by expansion in steel using industries that were strongly, though certainly not exclusively, connected to a booming market for consumer goods; above all, motor vehicles. Despite the loosening of the constraints on international trade, much of the increased output of steel within the old industrial countries was absorbed by growing steel-consuming industries located within their national territories. Indeed in some countries, state involvement in steel production increased, whether by closer private-public sector collaboration or by nationalization as the state took direct responsibility for steel production. Modernization of a key industry evidently could not be left to the private sector, not least as it often lay at the heart of more far-reaching industrial restructuring policies.

In other ways, though, the international geography of steel production was beginning to alter. For as more former colonies gained formal political independence, some of them embarked on post-colonial industrialization strategies while others that had achieved such independence at an earlier stage revived or strengthened plans for autonomous industrial development. Since the prime motive behind such plans was to break the economic dependency relations imposed in an earlier (formal or informal) imperial era, there was often heavy state support for embryonic steel producers since these were seen as central to evolving strategies of industrialization and modernization. But while global capacity and output grew in the 1960s, increasing attention came to be directed towards export markets from both old and newly industrializing countries; nowhere was this more evident than in Japan, following a phenomenal post-war resurgence of steel production there.

After 1973, of course, things took a markedly different turn. The long post-war boom stuttered to an end and, triggered by crude oil price rises, was transformed into a deep recession. The steel industry became characterized by global over-production crises, chronic over-capacity, especially of bulk production of ordinary steels, and plummeting profits or burgeoning losses from steel production. These problems were sharply exacerbated because of major expansion plans, in old and newly industrializing countries alike, formulated in response to extrapolations of the growth trends of the 1960s. The time-scale involved in constructing and commissioning new production complexes meant that yet more capacity was coming on-stream just as global demand collapsed. Partial and uneven recovery from recession — with another downturn precipitated by further oil price rises in 1979–80 — coupled with continuing chronic over-capacity in steel, resulted in periodic resurgences of protectionist sentiment and policies (especially in the USA). Heightened and new forms of state involvement in the steel industry emerged as a new and changing geography of international production and trade was created. More and more steel became traded internationally. The balance of production swung away from old production areas such as north-west Europe and the USA and towards Japan and a select few of the newly industrializing countries (NICs). In contrast to earlier periods, national states in the old industrial countries were more willing to countenance the decline of steel production within their national territory; some, as in France and the UK, positively encouraged and welcomed it. In part this was because of the volume of internationally traded steel and the variety of low-cost producers from whom steel could be imported but it also reflected the way in which steel no longer occupied so central a role in political and economic affairs (see Figure 1.1). This was especially evident in defence-related industries. The means of destruction had become technically more sophisticated; computers and micro-electronics came to be seen much more as the strategic industries rather than steel.

The changing intranational geography of production in the old steel-producing countries

While this changing international context is important, in and of itself it cannot explain why the geography of steel production changed as it did within countries such as France, West Germany, the UK, and the USA. It is also necessary to consider, in particular, the interrelated effects of new production technologies and changing sources of raw materials in influencing corporate and state strategies as to where to produce steel.

The early geography of iron and steel production in such countries was heavily constrained by the location of raw materials and the costs of transporting them (for example, see Hartshorne

Figure 1.1 'Decline of the steel barons' (*Financial Times*, 7 October 1985)

This week, the leaders of the world's major steel producers are meeting in London, but few people outside the industry will take much notice. The state of the world's steel industry is such that there is little likelihood of any great rows or dealmaking at this year's conference in London or, for that matter, for the foreseeable future. For the truth is that this once mighty industry has been largely stripped of its power.

There was a time, in 1962, when a chairman of the United States Steel Corporation, Roger Blough, could even take on a U.S. President, John F. Kennedy, over steel prices, and the world trembled. And it was not so long ago that car makers and other big users would wine and dine their steel suppliers just to try and get enough tonnage to keep going. But today, world steel overcapacity is so great and so deeply entrenched that buyers everywhere can sit back and play one steelman off against another with impunity.

It was also not so long ago that international trade in steel was a minor and quietly managed part of the business. The U.S. topped up its needs by importing from Canada, and left the Europeans to supply their former colonies and other developing countries. In 1950, only 20m tonnes of steel were traded internationally. Now, more than 140m tonnes a year move around the world, and there are dozens of new and aggressive producers battling for bigger shares in a pretty flat market. The result has been chaos in international markets, forcing governments everywhere to intervene with various types of production and import controls on steel.

But the discussions are mainly between politicians and bureaucrats. The once mighty steel barons are mainly spectators, waiting to be told under what conditions they must operate. They still have lots to do, but at a much more mundane level, struggling like any other business to cut costs and make better products.

1928; Martin 1957). Initially iron production was profoundly affected by the availability of charcoal and ores. As industrial capitalism established its grip over iron and then steel and changes in production technologies came about, the decisive influence became the location of coking coals and, to a lesser extent, iron ores. In one sense, these initial geographies were powerfully shaped by nature and the natural environment, by the locations of what could become resources, given contemporary production and transport technologies. As technologies changed and/or domestic reserves of raw materials became worked out, a series of pressures emerged to alter these initial spatial patterns; but counterpoised to them were other reasons not to alter these patterns. These latter arose from the costs of writing-off fixed capital in established, generally inland, production sites.

Such tensions were already evident in the nineteenth century. For example, changes in steel production technologies on occasion necessitated switching to imported ores as locally available ones were unsuitable for the latest techniques (for example, on Teesside in north-east England; see Hudson and Sadler 1985a). But a combination of the political and economic environment of the inter-war years, the imperatives of war-time production, and the fact that the decisive technological breakthroughs did not emerge until the 1950s meant that it was not until that decade that their full force was felt and the extent of their implications for the intra-national geography of steel production became apparent. The first of these major technological changes involved the introduction of basic oxygen steelmaking (BOS), cutting the socially necessary labour time to convert liquid iron to steel from around eight hours, using open hearth technologies, to around forty-five minutes. This was a fairly dramatic reduction and the adoption of BOS technology led in time to equally dramatic changes in the geography of steel production. In particular, it further emphasized the move to coastal locations, for two reasons. First, this facilitated imports of raw materials, especially richer iron ores to help satisfy the greatly increased demands for hot iron (since, on average, a BOS vessel operates on a charge of 70 per cent molten iron and 30 per cent scrap steel while the open hearth process utilized a higher proportion of cold inputs). Second, it allowed major modifications to be made to existing coastal or quasi-coastal plants or the construction of totally new ones, often mimicking the massive innovatory Japanese complexes which were setting fresh productivity norms, and centred around a tremendous expansion in hot iron-making capacity via the erection of much bigger blast furnaces. In some cases, such as France, Italy, or the UK, national governments seized this chance to reorganize the intranational geography of production to use the steel industry as an instrument of regional policy, concentrating investment in new capacity in peripheral problem regions.

But technological changes extended widely, affecting much more than just bulk liquid iron and steel production. The introduction of continuous casting (CONCAST) techniques, eliminating the need to cast semi-finished steel in ingots, greatly increased production efficiency and cut costs via reducing energy inputs and avoiding the need to reheat steel prior to rolling. Increasingly rolling mills become controlled by computer, allowing more precise specification of standards and less wastage in the process. Indeed, more generally the introduction of computer-control systems, greater automation and an increasing tendency to move towards production organized on a continuous flow-line,[2] intensified pressures to organize bulk steel production in major integrated complexes on green-field coastal sites. This was intended to allow rational technical planning, the realization of scale economies (provided such complexes were operated at high levels of capacity utilization), and sourcing via cheaper and/or higher grade raw materials.

But if all this was true of ordinary steels, it was not the case with respect to special steels production.[3] This involved a very different scale of both physical output and technologies: typically, small batch production of much higher value-added products, with steel produced in electric arc furnaces from scrap. The geography of special steels proved much less susceptible to change — although a proportion of the new coastal complexes such as Fos-sur-Mer, near Marseilles, did contain some special steels production facilities — than was the case with ordinary steel produced in bulk.

In any case, the coastal expansion of bulk production of ordinary steels was taking place at a time of strongly growing aggregate demand for steel so that the longer-term implications of this growth for smaller inland plants, using either open hearth or BOS technologies, were substantially masked in the early 1960s and 1970s. Even in those cases where coastal expansion would lead to some net loss of jobs, the seeming promise of secure long-term employment for most steelworkers defused trade union opposition to reorganization; indeed, steel unions were often enthusiastic supporters of plans to invest heavily in new capacity, even if this resulted in some alteration in the geographies of production and employment. The decisive imperatives were those of maximizing steel output; the emphasis was firmly established on capacity expansion, not capacity closures. When global steel demand slumped sharply after 1974 and the subsequent weak recovery in demand was accompanied by intensifying pressures of international competition from lower-cost producers, the future of the older inland plants became an increasingly precarious one. In due course, equally serious doubts were raised as to the viability of some of the new coastal complexes as the future for steelmaking in the old industrial countries began to appear a very bleak one. More and more, the agenda of political and public debate was dominated by discussions of capacity closure, job losses and the destruction of communities; of steel companies diversifying out of steel (for example into aluminium, micro-electronics, oil, or trading activities) and out of steel towns.

Steel, class, and community; industrialization, de-industrialization, and re-industrialization in the old steel towns

The initial phases of capitalist industrialization in countries such as the UK, USA, and West Germany witnessed the establishment of capitalist relations of production in the manufacture of iron and steel as one route to producing profits. A necessary condition for this to happen was that the required labour-power was assembled so as to be available for employment in the production process. This implied the creation of socialization mechanisms that would lead employees of the iron and steel companies routinely to accept their class position as wage labour, backed up by the coercive powers of the state and its monopoly over the legitimate use of violence should they choose to challenge it. It also implied provision of a built environment within which the biological and social reproduction of labour-power could occur; although the logic of the accumulation process dictated that this provision be at minimum expense since in this phase, in its entirety, it formed a direct cost of production to those private capitals producing steel.

Steel towns, then, were materially and socially constructed in the image of the dominant social relations of production; above all, they were both shaped by and helped to shape the class relationship between capital and labour. It would be wrong to see labour as a passive participant in forging this relationship. Workers were actively involved in the process of creating social divisions within the workforces of iron and steel plants, associated with the development of a particular occupational and technical division of labour within the production process (see Bowen 1976). These social divisions within the workplace in turn extended in their implications beyond the boundaries of the plant into community life, and into emergent local and national working-class political organizations. Both these workplace- and non-workplace-based social relations between capital and labour and those within the labour force itself permeated the construction and reproduction of communities. These were places based upon often relatively high male wages from employment in steel production, characterized by a particular gender division of labour centred around waged male work outside the home and unwaged female work within it. Sons followed in their fathers' and grandfathers' footsteps into 'the works'. Industrial relations too were cast in a deeply paternalistic mould. The new boom towns were places where, in a very real sense, life revolved around the steelworks, often the only major source of waged employment (see Figure 1.2).

Within these fairly precise limits, this satisfied the interests of

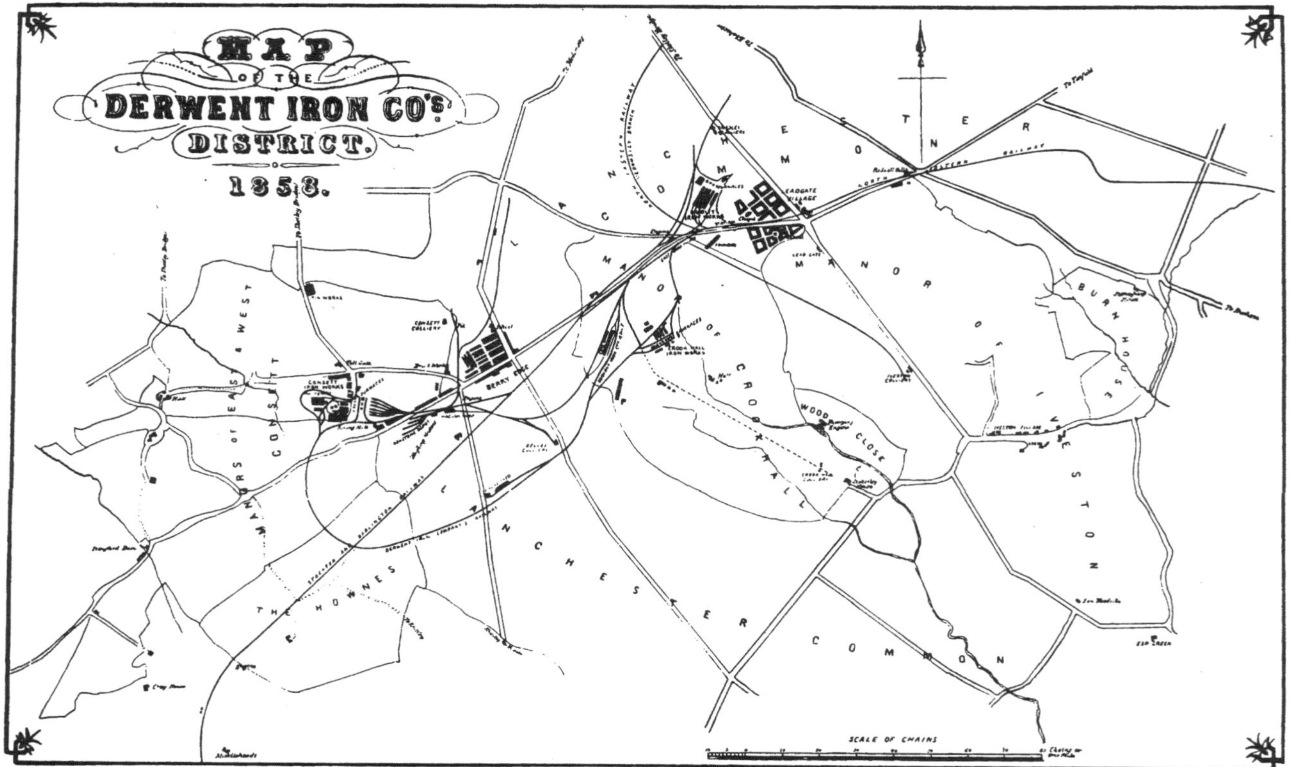

Figure 1.2 Map of the Derwent Iron Company's district, 1858

the major social actors, cyclical variations notwithstanding, as capitalists made profits while their workforce found waged employment. Social life in these places could thus go on and they could reproduce themselves socially for as long as the steelworks provided well-paid jobs. For the steel companies often jealously guarded their dominance in spatially defined male labour markets, actively preventing the introduction of alternative sources of male employment that would challenge their position as virtual monopoly buyers of male labour-power. At the same time, the combination of the character of the dominant working-class culture, its rigid gender division of labour, and the consequent constraints that the shift working patterns in the steelworks imposed on married women's ability to accept waged work, effectively meant that any diversification of opportunities for wage labour involving women was strictly limited. Once it became clear that the steel companies no longer required labour-power in the volumes that they had once done, however, the smooth social reproduction of these communities was placed in jeopardy. As it became evident in the 1970s that hundreds, maybe thousands, of jobs were to be lost and that entire works were to be closed, the economic rationale was ripped away from these places.

This was a truly traumatic process. Not least, because it was only too apparent that such changes were not simply a product of private capital disinvesting, bemoaning a lack of sufficient profits. For a variety of reasons, by the mid-1970s national states (and in the case of the European Community, an embryonic supranational state) had often become deeply involved in the organization of steel production, most visibly via nationalization or other forms of public ownership. Consequently they were very openly involved in decisions to destroy communities. In these circumstances, it was impossible simply to attribute mass unemployment, poverty, and the social costs of industrial decline to the decisions of private capitals guided via the 'hidden hand' of the market. The very visible hand of the state was only too apparent as a central element in the whole process. This transparently politicized closures. Along with the pronounced locational concentration of job losses, and, to some extent, the still-symbolic significance of steel as a measure of national industrial virility, it was instrumental in triggering a number of campaigns to contest closure proposals. Such campaigns were conducted in various ways — from attempts at rational argument, to rounds of public meetings and parliamentary lobbies, to riots on the streets. In the end, with very rare exceptions, they shared the same fate: failure.

In part this was because state policies over redundancy payments deliberately or inadvertently divided workforces within plants, as older workers were more inclined to take what seemed to them to be major sums of money whereas younger men wanted to fight for their jobs. Some were seduced by promises of alternative jobs. It also reflected divisions between trade unions in the steel industry, both within and between national states, that had developed over a long period. Occupational divisions and a certain territorial chauvinism combined to divide rather than unite steelworkers, let alone enable them to secure support from other social groups in their own communities and regions and other trade unionists and workers in other industries and locations. The net result was that it proved very difficult to construct and maintain a sufficiently broad basis of support to prevent anti-closure campaigns developing in such a way as to put one plant in competition with another, and steelworkers in one region or country in competition with steelworkers in another region or country.

The process of creating competition between groups of steelworkers, one in which they themselves were active participants, in the end simply made it easier for states and capitals to push through their proposals for restructuring via capacity closures and job losses. It thereby also enabled them to impose new working practices and conditions to increase labour productivity in those plants that did remain open. The old 'cosy relationships' between management and workers, born of an earlier era of sustained growth, were pulled apart as the terms on which a wage was offered in exchange for labour-power changed for the worse for workers as managers asserted what they saw as their right to manage.

Given that former steel towns had been de-industrialized, the question arose as to whether, and on what terms, they could be re-industrialized. Could new forms of state policies be devised to attract private capital to former steel towns to provide new sources and forms of wage labour? These re-industrialization measures were constructed in response to local demands for alternative jobs, partly because local politicians in steel closure areas felt the electoral and political necessity to respond to such demands and partly in response to more general pressures on states to legitimize their involvement in the destruction of steel employment in these towns in the first instance. Such policies have typically been conceived around the twin axes of small areas and small firms. Their content encompassed (and encompasses still, by and large) a familiar mixture of forms of activity: creating new general conditions of production via infrastructure investment plus grants, loans, and wage subsidies to private capitals, as ex-steel towns joined the competition to sell themselves in a global place market.

Often, as in places such as Consett, this led to a promotional hype that can obscure the extent and character of the problems to be tackled. This is all the more poignant when set against the job-creating impacts of such policies. In general, they have been conspicuously unsuccessful, creating few jobs, often on a temporary and precarious basis, with poor working conditions and low wages. Such jobs as have been created have often gone to the daughters and wives of ex-steelworkers, as the previous constraints on the gender division of waged and unwaged labour have been shattered by the cessation of steel production. This in itself is profoundly indicative of how social relations inside and outside the workplace have changed and are changing in former steel towns. Against a backdrop of mass unemployment, companies can selectively recruit 'green' female labour on their own terms as women seek to boost household incomes in the knowledge that the chances of their husbands and sons securing permanent well-paid jobs there are minimal.[4] Given the failure of re-industrialization policies to stimulate even a fraction of the required new jobs, other forms of state involvement have been developed to control and manage life in these ex-steel towns. These have involved the creation and expansion of temporary job and training schemes and welfare provision to ensure a sort of on-going existence for people in such places. In the final analysis, the effect, if not necessarily the intention, of such forms of activity is to persuade their inhabitants to come to terms with their lot as people who have, in all senses, become socially marginalized.

The organization of the book

In the chapters which follow we set out to build upon and explore some of the implications of these introductory remarks. In Chapter 2 we detail major new patterns which have emerged in steel production, trade, and consumption at a global scale. In particular, we consider the expansion of Japan and the two largest NIC steel producers, Brazil and South Korea, and counterpoise this against the steady decline of western Europe and the USA. We look also at the growing volume of steel traded internationally and the role of government in regulating this; at changing patterns of supply in the steel industry's key raw materials of coking coal, iron ore, and scrap, and at evolving trends in one of steel's major market sectors, motor vehicles.

In the next two chapters we address the process of change in particular localities in selected national states and at particular moments, using the pattern of global steel production as a context against which to view a range of complex and often contradictory developments. In Chapter 3 we consider how closure proposals have been contested, by and large unsuccessfully, in a number of different environments: at Consett, Corby, and Ravenscraig in the UK; in the Alsace-Lorraine region of France; in the Ruhr valley in West Germany; and in other European Community member states and in the USA. We analyse why such anti-closure campaigns developed, why they assumed the forms that they did, and why they met with such little success.

In Chapter 4 we switch attention to the measures adopted in response to steel closures and to the demands posed in anti-closure campaigns. These are often referred to as re-industrialization policies. We look at their effectiveness in terms both of their own stated objectives and their economic and social impacts in steel closure areas, drawing on the subtly differing examples of Consett, Corby, Hartlepool, and Sheffield in the UK, and the Nord-Pas de Calais region in France. We also examine a range of similar policy responses developed within the institutions of the European Community. Finally, we evaluate the longer-term social and political implications of such national and supranational measures. This forms the point of departure for our concluding chapter, in which we discuss some of the political and theoretical implications of the developments analyzed in the book and consider some possible future directions in both spheres.

Notes

1 These have been referred to as coal combines in the UK and in West Germany as the *montanindustrier* (see Hudson 1989).

2 There are those (for example Scott and Storper 1989) who see the steel industry as organized on Fordist lines. Ignoring for the moment the fact that special steels continue to be produced on a small batch basis, to conflate continuous production processes in the bulk manufacture of ordinary steel with the Fordist labour process and, more generally, to equate all mass production methods with Fordism robs the latter of its analytic precision and value.

3 At a very minimum a distinction must be drawn between ordinary steels and special steels. 'Special steels' refers to higher grade, higher value-added steels, to which various alloys or combinations of alloys have been added to give them specific chemical and physical properties. However, there is considerable heterogeneity within both broad categories and recent technological developments in production processes have further blurred the distinction between them (see Hudson and Sadler 1987b).

4 Lest there be any doubt about it, we emphasize that we are not arguing

for ‘male’ rather than ‘female’ jobs; rather the point is to stress the conditions in which women have now become ‘free’ to seek waged labour in places where labour market conditions for them are extremely unfavourable.

Chapter two

New patterns of production and trade in the world steel industry

Introduction: the steel industry and the geography of international economic activity

For most of the twentieth century, the world's steel industry knew only one path: expansion. Led at first by the UK, the USA, and Germany, total world steel production followed a steadily rising trend in the first decades of the century, weathering destruction of capacity in the course of two global conflicts and an inter-war economic slump to return with stronger annual growth rates. After 1946 world steel output grew even more dramatically, in response to strong demand from dynamic steel-consuming sectors such as motor vehicles. From 1946 to 1974 total world steel output increased more than six-fold, from 112 to 704 m tonnes (Table 2.1). In only three years during this period was there less steel produced than in the twelve months immediately preceding — 1954, 1958, and 1971 — and in each case the drop was more than compensated for by an increase the following year. Small wonder, then, that by the end of the 1960s steel producers envisaged a prospect of enduring world production growth. One speaker at the 1973 meeting of the International Iron and Steel Institute saw the chief limiting factor on strong worldwide demand as 'the ability of the industry to produce' (quoted in Hogan 1983: 15). Clearly yet more steelmaking plant was seen to be needed. Total new capacity planned in the western world alone in 1974 amounted to 240 m tonnes, with the bulk of this expansion expected to come from western Europe and south-east Asia, and smaller yet significant increases in Latin America, north America, and the Middle East (see Table 2.2). All regions of the world, it seemed, needed more steel.

In the event, however, 1974 marked a turning-point of a very different kind. The world's capitalist economies collectively plunged into a recession triggered by OPEC oil price increases,

Table 2.1 World crude steel production, 1900–83 (m tonnes)

1900	28.3
1910	60.3
1920	72.5
1930	95.1
1940	140.6
1945	113.1
1950	191.6
1955	270.0
1960	346.4
1965	454.0
1970	595.4
1974	703.5
1975	643.4
1976	675.4
1977	675.5
1978	716.9
1979	746.7
1980	716.2
1981	707.7
1982	645.2
1983	663.4

Source: International Iron and Steel Institute

Table 2.2 Planned new steelmaking capacity, 1974 (m tonnes)

Western Europe	68.0
(of which EC	41.3)
South-east Asia	67.9
Latin America	37.2
North America	28.5
Middle East	23.8
Africa	12.3
Oceania	2.3
Total	240.0

Source: Hogan 1983: 17

clearly revealing a fragility in the long post-war boom which had in fact been growing since the late 1960s. High levels of steel stocks had been built up by consumers during 1973–74 in anticipation of threatened shortages. When downturn came, these proved to be surplus to requirements, and served to depress still further the level of world steel demand in 1975. World output slumped from the previous year's total by nearly 10 per cent and did not recover that level again until 1978. Demand picked up in 1979 to sustain a new record output of 747 m tonnes; but this was again to prove only another temporary peak as production slumped in each of the following three years, to reach 645 m tonnes by 1982. On only two previous occasions in the century — the great Depression (1930–32) and the rundown of war economies (1944–46) — had annual output dropped consecutively over a three-year period.

Table 2.3 The geographical distribution of world steel production changes, 1974–84

	m tonnes		*% change*
	1974	*1984*	*1974–84*
Industrialized countries	463	376	−19
Developing countries	31	70	+126
Centrally planned economies	210	264	+26
Total	704	710	+1

Source: NEDO 1986a

In the new global economic climate, divergent patterns of growth in steel production in different areas became increasingly significant. The developed economies of western Europe, north America, and Japan entered a period of protracted decline, while the newly industrializing countries, most especially in south-east Asia and Latin America, saw a no less dramatic increase (Table 2.3). By 1984, after two cycles of decline and recovery, total world steel output was at almost exactly the same level as in 1974; but there had been a 19 per cent drop in output in the developed economies (to 376 m tonnes) and a 126 per cent increase in the newly industrializing economies (to 70 m tonnes). This emerging pattern was due to a number of factors. Steel consumption traditionally grows most strongly in the early stages of industrialization, before the growth rate declines in later stages. Within the developed economies after 1974 steel was increasingly vulnerable to substitution by other products such as plastics or aluminium. These were less energy-intensive at the production stage, lowering overall costs significantly at a time of high energy prices; and/or lighter, giving rise to product improvements such as increased fuel efficiency in motor vehicles. Within the newly industrializing countries (NICs), on the other hand, production was growing in response to the demands of early stages of planned industrialization programmes and the ready availability of steel production machinery and technology, often supplied through subsidiary companies of established European or American steel producers. Such trends had in fact already been apparent, but their significance was heightened after

1974 as ambitious expansion programmes in both developed and newly industrializing economies, formulated in response to strong growth forecasts, came to full or partial fruition, exacerbating already severe problems of over-capacity and geographical imbalance.

Table 2.4 World crude steel production and consumption, 1984 (m tonnes)

	Production	*Consumption*	*Balance*
Japan	106	74	32
European Community	120	95	25
Other industrialized economies	64	56	8
Newly industrializing economies	70	105	−35

Source: NEDO 1986a

The current and prospective pattern of world steel production and consumption largely reflects these trends (see Table 2.4). Both the European Community and Japan produce more steel than they consume, while the United States market is heavily penetrated by imports: a major factor in the growth of protectionist measures between these trading partners. The total market in developing countries, on the other hand, remains substantially greater than production, although ambitious expansion plans abound. The key trade battle, then, is currently being fought between Japanese, European and NIC producers for the growth markets of other NICs. This is being waged in a context where most forecasts only envisage marginal total world demand increases for the short to medium term, and where considerable problems of over-capacity still remain in countries of the developed world prevented from exporting their surplus by trade restrictions (see NEDO 1986a: 26–31).

Within this chapter we seek to expand upon these remarks by considering three dimensions to these new patterns of international steel production and trade. We examine the picture of output growth in the NICs and the changing fortunes of the steel industry in Japan, and compare this to the almost unremitting prospect of decline in western Europe and the USA.[1] We move on to consider how the geographical balance of production has been mediated by the impact of international trade, and the growing role of government policies in regulating this trade. We conclude by reviewing the changing patterns of supply of the steel industry's raw materials, and the changing global distribution of the key steel-consuming sector of motor vehicles. In short, we locate the new geography of the world steel industry in the context of the changing global geography of industrial production.

Faltering or continuing growth? The NICs and Japan

Japan

In the period after 1945 the Japanese steel industry saw the most remarkable growth of any steel industry in history. Its output soared from 1 m tonnes in 1947 to 22 m tonnes in 1960 and 93 m tonnes in 1970, and reached 119 m tonnes in 1973. This supplied both domestic demand and a growing world export market, and was underpinned by a massive programme of support from the country's main financial institutions. New capacity was constructed in large coastal complexes with massive blast furnaces. Ten sites each had an annual capacity in excess of 8 m tonnes while the western world's largest, at Fukuyama City, had an annual capacity of 16 m tonnes (Hogan 1983: 65). In 1985 Japan was the western world's largest steel producer, exceeded only by the USSR, and its largest steel company, Nippon Steel, dwarfed all other producers with an output of 28.6 m tonnes. Five of the world's top eighteen steel companies were Japanese (see Tables 2.5 and 2.6). Yet spectacular though this growth has been, and while many NICs now attempt to model their industrial strategies on Japan's export-led performance, the Japanese steel industry currently faces a traumatic period of crisis and rationalization.

There should be no doubt that the success of the Japanese steel industry was based upon exports. Dramatic growth in the 1960s was expressly designed to feed world markets. The construction of steelworks on deep-water harbours was essential both for the import of raw materials and the export of steel. In 1960 the Japanese steel industry exported 14 per cent of its output; by 1975 exports accounted for 34 per cent of production (Table 2.7). To this growth in the volume of direct steel trade should also be added the proportion of steel traded indirectly, in the form for example of ships, machinery, motor vehicles, or other consumer goods. On this basis one estimate is that at least half of Japan's total steel output was ultimately destined for export (Hogan 1983: 83).

Within this dependence upon export markets lay the seeds of the Japanese steel industry's current crisis. Just as in other advanced economies, Japanese steel companies initiated ambitious expansion programmes in 1973 and 1974 (on the eve of world recession) to add a further 22 m tonnes of capacity. These plans, entailing the construction of seven new blast furnaces (of which only two were

Table 2.5 World steel output, 1985 (m tonnes)

USSR	155.2
Japan	105.2
USA	80.4
China	46.5
West Germany	40.5
Italy	23.7
Brazil	20.5
France	18.8
Poland	16.1
UK	15.7
Czechoslovakia	15.2
Canada	14.7
Romania	14.4
Spain	14.2
South Korea	13.5
India	11.1
Belgium	10.7
South Africa	8.6
North Korea	8.4
East Germany	7.9
Mexico	7.3
Australia	6.4
Netherlands	5.5
Taiwan	5.1
Turkey	5.0
Sweden	4.8
Austria	4.7
Yugoslavia	4.4
Luxembourg	3.9
Hungary	3.7
Bulgaria	3.0
Venezuela	3.0
Argentina	2.9
Finland	2.5
Other countries	16.4
World total	719.9

Source: International Iron and Steel Institute

considered as replacement for older plant), were completed by 1978. They also included Nippon Kokan's ten-year scheme to construct a wholly new steel plant on an artificial island at Ohgishima (see Junkerman 1987). Yet as the recession intensified, key export markets such as the USA retreated behind trade barriers, with protection the new order of the day. Former NIC markets such as Brazil and South Korea were increasingly sourced from domestic production, not Japanese exports. And when other sectors felt the brunt of recession and trade restriction, Japanese domestic steel demand dropped as well.

Table 2.6 Major steel companies, 1985 (all companies producing 2 m tonnes or more)

	Ranking	*Output (m tonnes)*
Nippon Steel (Japan)	1	28.56
US Steel (USA)	2	15.15
Finsider (Italy)	3	13.45
BSC (UK)	4	13.35
Siderbras (Brazil)	5	13.23
NKK (Japan)	6	12.10
LTV (USA)	7	11.90
Thyssen (Germany)	8	11.07
Sumitomo (Japan)	9	10.99
Arbed Group (Luxembourg)	10	10.97
Kawasaki (Japan)	11	10.86
Bethlehem (USA)	12	9.47
Posco (S. Korea)	13	9.26
Sacilor (France)	14	8.75
Usinor (France)	15	7.22
Sail (India)	16	6.82
Iscor (South Africa)	17	6.82
Kobe (Japan)	18	6.46
BHP (Australia)	19	6.32
Inland (USA)	20	5.51
Hooghovens (Netherlands)	21	5.49
Armco (USA)	22	4.85
Klöckner (Germany)	23	4.59*
Stelco (Canada)	24	4.53
Cockerill-Sambre (Belgium)	25	4.50
Voest Alpine group (Austria)	26	4.46
Mannesmann (Germany)	27	4.40
Ensidesa (Spain)	28	4.40
National Steel (USA)	29	4.35
Dofasco (Canada)	30	4.37
Krupp (Germany)	31	4.22*
Sidermex (Mexico)	32	4.16
Hoesch (Germany)	33	4.10
Peine-Salzgitter (Germany)	34	3.82
Nisshin (Japan)	35	3.29
China Steel (Taiwan)	36	3.20
SSAB (Sweden)	37	3.01
Tokyo Steel (Japan)	38	2.78
Sidor (Venezuela)	39	2.72
Wheeling-Pittsburgh (USA)	40	2.54
Rouge Steel Co. (USA)	41	2.50
Algoma (Canada)	42	2.45
Weirton (USA)	43	2.41
Tata Iron & Steel (India)	44	2.06

Source: NEDO 1986a

Note: *Financial year to 30 September

Table 2.7 Japanese steel production and exports, 1960–80

	Production (m tonnes)	*Exports* (m tonnes)*	*% of production exported*
1960	22.1	3.1	14
1965	41.2	12.7	31
1970	93.3	22.3	24
1975	102.3	34.4	34
1980	111.4	34.1	31

Source: Hogan 1983: 82–3

Note: *Crude steel equivalent

In response, Japanese steel companies adopted a number of new strategies in the 1980s. All the major companies have diversified into the manufacture of silicon wafers for the electronics industry. Kawasaki Steel anticipates that its non-steel business, centred on electronics, will generate 40 per cent of its income by the end of the century (compared to 20 per cent in 1985), while Nippon Steel estimates that electronics will provide 20 per cent of its turnover by 1995. The steel companies did not forsake their primary business activity, though, continuing with modernization investments to reduce costs and improve product quality. Partly as a defensive reaction against import quotas, some companies also invested in steel production overseas; for example in 1984 Nippon Kokan took a 50 per cent stake in National Steel, then the sixth largest US steel company.

Yet even these measures did not prove sufficient to restore and maintain balance sheet profitability. In the half-year to September 1986 the five major companies sustained losses of £790 million. Within a matter of months they had all announced restructuring plans, aimed at reducing their combined workforce by 44,000 to 132,000 by 1990 (Table 2.8). They were based on a forecast

Table 2.8 Japanese steel companies' employment plans, 1987–90

	1987 employment	*Forecast 1990 employment*
Nippon Steel	65,000	45,000
Nippon Kokan	30,000	23,000
Kobe Steel	28,000	22,000
Sumitomo Metal	27,000	21,000
Kawasaki Steel	26,000	21,000
	176,000	132,000

Source: Financial Times, 28 February 1987

national output in that year of 90 m tonnes against 96 m tonnes in 1986, and included some dramatic cut-backs. The biggest company, Nippon Steel, aimed to slash its annual capacity from 34 m tonnes to 24 m tonnes. Kawasaki Steel announced that it was contemplating the total closure of one of its two works. In keeping with the tenets of Japanese employment philosophy, most of the workforce reductions were to be undertaken through transfer within the company to other sectors or through early retirement. But the key question which remained was whether these cutbacks would prove to be enough to restore profitability.

Brazil

In 1970 Brazil was the nineteenth largest steel producer in the world. By the mid-1980s, after a decade of growth which had seen output increase three-fold to 21 m tonnes, the Brazilian steel industry was the largest of any newly industrializing country, and the seventh largest in the world. Siderbras, the state-owned producer, had become the globe's fifth largest steel company (see Tables 2.5, 2.6, and 2.9). Yet this government-backed programme of expansion had run into a number of problems. It was conceived at a time of rising domestic demand in the 1970s, in the context of a general industrial growth strategy. Downturn in the economy in the early 1980s effectively wiped out several years' growth and forced the expanding steel industry to concentrate more and more upon export markets. This, coupled with construction delays, cost overruns, currency fluctuations, and government control of some domestic prices from 1979 to 1985, forced Siderbras into growing indebtedness. At the same time the export markets upon which the industry had begun to depend came under a growing threat from trade protectionism. Together, these trends spelt considerable difficulty for the Brazilian steel industry on the eve of the 1990s.

The plans of the mid-1970s were formulated in the light of Brazil's 4 m tonne trade deficit in steel production in 1974. The five main producer companies within the Siderbras group were the result of this unprecedented expansion. They included Usiminas, in Minas Gerais, and Cosipa in Sao Paulo State, each with an annual output of 3.5 m tonnes; CSN's Volta Redonda mill; and the two newest works, Tubarao (at Vitoria) and Acominas (in Minas Gerais).

Tubarao was inaugurated in 1983, for an investment cost of 3.1 billion dollars and with an annual capacity of 3.2 m tonnes. It represented a first for Brazil: a steel plant designed specifically for export, in this case of semi-finished slabs. In recognition of the

Table 2.9 Brazilian steel production and trade, 1972–84 (m tonnes)

	Production		*Exports*	*Imports*	*Balance*
	Crude steel	*Rolled products*	*(product m tonnes)*		
1972	6.5	5.1	0.4	1.1	−0.7
1973	7.1	5.5	0.4	1.8	−1.2
1974	7.5	5.7	0.2	4.2	−4.0
1975	8.3	6.4	0.1	2.9	−2.8
1976	9.2	6.9	0.3	1.1	−0.8
1977	11.2	8.4	0.4	0.9	−0.5
1978	12.1	9.5	0.9	0.7	0.2
1979	13.9	10.0	1.5	0.6	0.9
1980	15.3	11.5	1.5	0.7	0.8
1981	13.2	9.8	1.9	0.9	1.0
1982	13.0	10.2	2.4	0.4	2.0
1983	14.7	11.1	5.2	0.1	5.1
1984	18.4				

Source: International Iron and Steel Institute

nature of its market, the company is an international conglomerate. Siderbras holds a controlling 51 per cent stake but two foreign partners, Kawasaki Steel of Japan and Finsider of Italy, hold the remaining 49 per cent, and are each obliged to market one-fifth of its output. By way of contrast, Acominas was originally conceived in 1976 as a plant to supply the domestic market, with an initial annual capacity of 2 m tonnes, to be built up to 10 m tonnes at a later date. Falling demand and major construction delays meant that the first stage was not completed until 1986. At a cost of 6 billion dollars it was some 2.5 billion dollars over the estimated budget. Expansion beyond this point has been postponed indefinitely and its output has been completely re-orientated, away from domestic sales and towards the export market.

Over-capacity and the attendant project delays at Acominas and elsewhere have been partially responsible for the growing financial problems of Siderbras. By 1986 its total debt amounted to 15.6 billion dollars, including 6.8 billion dollars owed to foreign banks, 7.5 billion to the Brazilian government, and 1.3 billion in short-term bank loans. As capacity expanded out of step with domestic demand, and in an attempt to meet the debt service burden, Siderbras turned more and more towards exports. By 1984 Brazil exported nearly 7 m tonnes of steel products, a far cry from the 4 m tonnes imported in 1974. Major export markets included the USA, China, and Japan (see Table 2.10). Yet a dramatic surge in exports after 1981 led the country into another series of problems, expecially in its trading relations with the USA. A number of anti-

Table 2.10 Brazilian steel trade by geographical region and by country, 1984 (m product tonnes)

By region

	Exports	*Imports*
European Community	0.41	0.05
Other western Europe	0.44	0.01
Eastern Europe	0.12	—
Africa	0.52	—
North America	1.46	—
Latin America	1.03	—
Middle East	0.44	—
South-east Asia	2.41	0.01
Oceania	0.03	—
Total	6.87	0.07

Major export markets by country (all export markets greater than 0.1m tonnes)

USA	1.34
China	0.71
Japan	0.61
Argentina	0.50
Algeria	0.43
Singapore	0.39
Ecuador	0.19
West Germany	0.19
Saudi Arabia	0.19
South Korea	0.19
Spain	0.18
Thailand	0.17
Turkey	0.17
Canada	0.12
Iran	0.12
Malaysia	0.12
USSR	0.12

Source: International Iron and Steel Institute

dumping complaints were lodged by US steel producers, plunging Brazil into the middle of a steel trade crisis. This was temporarily resolved at the end of 1984, when a five-year export restraint agreement was concluded, limiting Brazilian exports to 0.8 per cent of the US market for finished steel (approximately 0.8 m tonnes annually). A further 0.7 m tonnes of semi-finished steel slab exports annually were permitted (rising to 1.1 m tonnes over a fifteen-year period) from Tubarao, to supply the newly re-activated California Steel, owned 25 per cent by the Brazilian government and 25 per cent by Kawasaki Steel of Japan (which partly owned

Tubarao); producing a graphic example of the growing internationalization of steel production and marketing arrangements as a means of at least partially overcoming international trade sanctions. But whether this would be sufficient to resolve the vast problems of Brazil's over-capacity remained to be seen.

South Korea

The second largest steel industry among NICs is that of South Korea. An output of 13.5 m tonnes in 1985 made it the fifteenth largest producer country in the world. The state-backed company, Pohang Iron and Steel (Posco), produced 9.3 m tonnes of this total, making it the thirteenth largest steel company in the world (see Tables 2.5, 2.6, and 2.11). Just as in Brazil, this prominence was the result of a conscious growth strategy adopted during the 1970s in an attempt to foster domestic industry and reduce dependence on imports. Unlike Brazil, however, South Korea has no reserves of iron ore, making it more akin to Japan in its complete dependence upon imported raw materials. The two countries also differ in their trade patterns and in the significance attached to exports, demonstrating the heterogeneity within a general pattern of NIC steel producer growth.

Table 2.11 South Korean steel production and trade, 1976–84 (m tonnes)

	Production		*Exports*	*Imports*	*Balance*
	Crude steel	*Rolled products*	*(product m tonnes)*		
1976	3.5	3.8	1.3	1.7	−0.4
1977	4.3	5.0	1.3	2.2	−0.9
1978	5.0	6.4	1.6	2.9	−1.3
1979	7.6	7.6	3.1	2.6	0.5
1980	8.6	7.9	4.5	1.9	2.6
1981	10.8	9.7	4.7	2.0	2.7
1982	11.8	11.0	5.5	1.3	3.2
1983	11.9	11.8	5.7	2.1	3.6
1984	13.0				

Source: International Iron and Steel Institute

The major works of the state company, Posco, was the result of an ambitious expansion programme launched in 1970 and completed in 1981, at a cost of 3.3 billion dollars. Throughout this period the project was a major national priority. It was finished ahead of schedule and on budget, able to meet its central aim — to supply domestic demand at prices well below those prevailing for imports. By 1979 South Korean exports exceeded imports. As the

Table 2.12 South Korean steel trade by geographical region and by country, 1984 (m product tonnes)

By region

	Exports	*Imports*
European Community	0.01	0.17
Other western Europe	—	0.06
Eastern Europe	—	—
Africa	0.02	0.03
North America	1.93	0.01
Latin America	0.06	0.19
Middle East	1.20	—
South-east Asia	2.47	2.78
Oceania	0.10	0.05
Total	5.89	3.46

Major trade flows by country (all flows greater than 0.1 m tonnes)

	Exports		*Imports*
USA	1.86	Japan	2.69
Japan	1.50	Brazil	0.19
Saudi Arabia	1.03		
Philippines	0.22		
Indonesia	0.15		
Malaysia	0.14		
India	0.12		
Thailand	0.11		

Source: International Iron and Steel Institute

1980s progressed the company came to rely increasingly upon these export markets as a means of utilizing capacity. Domestic demand did not fall to the same extent as in Brazil, though, ensuring that a planned second plant, on the south coast at Kwangyang, with an additional capacity of 2.8 m tonnes, was authorized in the mid-1980s.

Like Brazil, a considerable proportion of South Korean exports in the mid-1980s was sent to the USA and Japan. In addition over 1 m tonnes was exported to Saudi Arabia in 1984; but China did not figure as a major export market, in contrast to the 0.7 m tonnes exported there from Brazil in the same year. South Korea also continued to import steel, again in contrast to Brazil, mostly from Japan (Table 2.12). In part this reflected the relative openness of its steel market, in part it was the result of political relations with Japan, the country on which its development path is partly modelled. This is especially true in the context of raw material supply, where Posco has followed the Japanese example by securing access to coking coal, in this case through its purchase of the Tanoma mine at Pennsylvania in the USA in the early 1980s.

In their different ways, then, South Korea and Brazil are illustrative of a number of tendencies amongst the NIC steel producers. They are now major producers, able to supply rising domestic demand and keen to expand an export market in both the developed world (especially Japan and the USA) and other NICs. The countries of the south-east Asian Pacific Rim figure prominently among their smaller yet significant steel trade flows (see Tables 2.10 and 2.12). Yet these two producer countries demonstrate the uncertainties as well as the opportunities of the global steel industry, and the problems of attempting to follow the 'Japanese model' in a changed economic climate, some two or three decades on. They are both acutely sensitive to domestic economic performance and to that of the world's economy at large, through the threat of trade sanctions within the European, American, and Japanese markets; and to the competitive threat in other NIC markets posed by the diversion of surplus developed country output as an indirect result of the same trade sanctions. The path of NIC expansion is a far from even or secure one.

Decline in Europe and the USA

The recent performance of the steel industry in western Europe and north America is, by strong contrast, one of almost unremitting decline. After 1974 demand and output slumped, profits turned to losses and redundancy and retrenchment came to dominate the boardroom agenda. These problems increasingly drew in national governments and within the European Community at least, also involved a degree of supranational policy formation in an attempt to resolve the deepening over-production crisis. In this section we outline some of the main production trends of the European Community (EC), the major non-EC producer states, and the USA.

The European Community

The steel industry has featured prominently since 1951 when France, West Germany, Belgium, Luxembourg, Italy, and the Netherlands founded the European Coal and Steel Community (ECSC). One of the central tasks in its early, formative years was to oversee post-war reconstruction of these two basic industries. As national economies boomed during the growth years of the 1950s and 1960s, and demand for steel seemed to be on an inevitably upward path, the activities of the ECSC and its eventual successor, the European Community, at least as far as steel was concerned, were confined to a supervisory role over the construction of new and replacement capacity. It was also able to offer more concrete support for investment programmes in new production capacity, acting as an intermediary in the procurement of low-interest loans on the global currency market, able to offer better terms to individual firms by virtue of its greater credit-worthiness as an institution.[2] With the onset of crisis in steel production from the early 1970s, however, the European Community was faced with a new problem of policy formation, particularly acute in that it had itself played a role in financing the development of growing over-capacity. It gradually expanded the scope of its measures, beginning to employ powers available since 1951 which had previously remained unused. By the early 1980s steel was distinctive amongst manufacturing industries in the extent and sophistication of regulation exercised by the European Community.

This system of control over steel production depended upon a shifting balance of power. The Commission of the EC, while appointed by member states, seeks to further the Community's interests as a whole. The Council of Ministers, on the other hand, consists of representatives from member state governments charged with securing the best deal for their country, and ultimately has the decisive voice. The possible range of points of conflict towards an industry like steel, in which national state governments were so integrally involved, was great amongst the original six-member Community. With the enlargement of the EC to incorporate the UK, Denmark, and Ireland in 1973, then Greece in 1981, and Spain and Portugal in 1986, the scope was far wider. The system of regulation has therefore entailed negotiation of a number of often delicate compromises between member state governments, major private and public sector steel producers, and the Commission.

These compromises have been made possible by the extent of decline in the EC's steel industry. Within the nine member states (as at 1973) total output slumped from 156 m tonnes in 1974 to 124 m tonnes the following year; climbed uncertainly back to 140 m tonnes by 1979; then slumped dramatically twice again to 128 m tonnes in 1980 and 110 m tonnes in 1982. Even after a slight recovery during 1984 and 1985, output remained some 20 per cent below the peak 1974 level (see Table 2.13). As new investment projects came on stream, the degree of over-capacity in the Community steel industry deepened. Capacity utilization rates dropped from a profitable 87 per cent in 1974 to around 65 per cent for the

Table 2.13 European Community crude steel production by member states, 1974–85 (m tonnes)

	1974	*1975*	*1976*	*1977*	*1978*	*1979*	*1980*	*1981*	*1982*	*1983*	*1984*	*1985*
West Germany	53.2	40.4	42.4	39.0	41.2	46.0	43.8	41.6	35.9	35.7	39.4	40.6
France	27.0	21.5	23.2	22.1	22.8	23.4	23.2	21.2	18.4	17.6	18.8	18.6
Italy	23.8	21.8	23.4	23.3	24.3	24.3	26.5	24.8	24.0	21.8	24.1	23.9
Netherlands	5.8	4.8	5.2	4.9	5.6	5.8	5.3	5.5	4.4	4.5	5.7	5.5
Belgium	16.2	11.6	12.1	11.3	12.6	13.4	12.3	12.3	10.0	10.2	11.3	10.7
Luxembourg	6.4	4.6	4.6	4.3	4.8	5.0	4.6	3.8	3.5	3.3	4.0	3.9
UK	22.4	19.8	22.4	20.5	20.3	21.5	11.3	15.3	13.7	15.0	15.2	15.8
Ireland	0.1	0.1	—	—	—	—	—	—	—	0.1	0.2	0.2
Denmark	0.5	0.6	0.7	0.7	0.9	0.8	0.7	0.6	0.6	0.5	0.5	0.5
Total	155.6	125.2	134.2	126.1	132.6	140.2	127.7	125.2	110.5	108.6	119.2	119.6

Source: Commission of the European Communities

Table 2.14 European Community steel trade, 1974–83

(External trade, m product tonnes, from and to the nine members as at 1973)

	Exports	*Imports*	*Balance*
1974	33.1	6.4	26.7
1975	27.6	7.1	20.5
1976	23.2	11.0	12.2
1977	27.7	11.0	16.7
1978	33.6	10.1	23.5
1979	31.9	10.8	21.1
1980	28.9	10.5	18.4
1981	32.5	7.6	24.9
1982	26.3	9.9	16.4
1983	27.4	9.8	17.6

Source: OECD

rest of the 1970s, falling even further to 57 per cent in 1982. The picture of decline was an almost constant one. Only in Italy did steel output hold up, in response to government policies (and often, allegedly, in defiance of European Community policy). Elsewhere output, capacity, and employment were slashed dramatically as losses rose and governments were increasingly called upon to come to the rescue through financial support or outright nationalization. The EC also represented a major force in world trade, not only in terms of the volume of steel traded within its boundaries, but also externally (see Table 2.14). Throughout this period, the Community consistently recorded a substantial trade surplus with the rest of the world, as producers sought to export the products of their excess capacity away from declining home markets.

In response to this deepening recession the Community developed a series of measures aimed at facilitating a relatively orderly restructuring amongst steel companies. These were first introduced in 1977 when the Commission approved a voluntary system of suggested minimum prices for certain steel products, which later became compulsory. During 1980 it became apparent that these measures were insufficient. The Commission was in practice unable to maintain or check upon price levels and a widespread practice of price discounting through, for example, the deliberate invocation of late-delivery clauses, had developed. During the second and third quarters of 1980 demand fell particularly sharply. Capacity utilization fell from 70 per cent in the first quarter to 58 per cent in the second, and in July orders were down by 16 per cent on the previous year.

While the Commission could impose minimum prices in its own

right under the Treaty of Paris (which established the ECSC), it could not implement stronger measures without approval from the Council. In particular it required the Council's consent to invoke Article 58 of the Treaty of Paris:

> In case of a decline in demand, if the High Authority deems that the Community is faced with a period of manifest crisis and that the means of action provided for in Article 57 (minimum prices) are not sufficient to cope with that situation it shall, after consulting the Consultative Committee and with the concurrence of the Council, establish a system of production quotas.

Historically the Council had been extremely reluctant to endorse any declaration of 'manifest crisis', refusing to do so even during a similar phase in the decline of the coal industry during the 1960s. On this occasion, however, so great was the problem, and so delicate were the negotiations arising from the pressures on most major European steel producers and their governments, that even the West German delegation (the most obdurate opponents of further state interference) agreed to support the declaration, after securing an agreement to exempt certain special steels from the production quota regime.

Under the quota system the Commission allocated production to each steel company on the basis of an average output from the preceding three-year period, in the light of short- and long-term forecasts of demand. It had the power to inspect and monitor production levels and could levy fines on producers which exceeded their quota (see Grabitz and Hanlon 1984). In 1981, to back up this system of short-range controls, the Commission sought and received the Council's approval for a further series of measures, giving it the ability to order capacity closures by effectively linking Community approval of state aid to steel producers with Community plans for the reduction of production capacity. The ultimate intention was to eliminate state aid altogether. A programme was agreed whereby national state aid and restructuring plans were to be notified to the Commission by 30 September 1982 and the Commission was to assess the proposals by 1 July 1983. All plans were to be phased so as to eliminate the necessity for state aid after 31 December 1985.[3]

By July 1983 substantial capacity reductions had been made by West Germany, France, and the UK. The Commission assessment called for much greater reduction before 1985 by Italy, but also substantial further cutbacks in most member states amounting to 8 m tonnes on top of the 16 m tonnes closed to that date (Table 2.15). At the same time it reaffirmed a determination to see state

Table 2.15 Commission proposals for capacity reductions

	Output capacity 1980 (m tonnes)	*Closures 1980–83 (m tonnes)*	*Further closures called for in July 1983 (m tonnes)*	*Total closures called for (m tonnes)*	*Total closures called for as a % of 1980 capacity*	*Closures 1980 to March 1985 (m tonnes)*	*Closures by March 1985 as a % of 1980 capacity*
West Germany	53.1	4.8	1.2	6.0	11.3	6.3	11.9
France	26.9	4.7	0.6	5.3	19.7	4.5	16.7
Italy	36.3	2.4	3.5	5.8	16.1	5.7	15.7
Netherlands	7.3	0.3	0.7	1.0	13.0	0.8	11.0
Belgium	16.0	1.7	1.4	3.1	19.4	3.1	19.4
Luxembourg	5.2	0.6	0.4	1.0	19.2	1.0	19.2
United Kingdom	22.8	4.0	0.5	4.5	19.7	4.6	20.2
Ireland	0.1	—	—	—	—	—	—
Denmark	0.9	0.0	—	0.0	7.0	ND	ND
Total	168.6	18.4	8.3	26.7	15.8	26.0	15.5

Source: European Commission

aid ended after 1985 in the context of specific restructuring plans:

> In essence, aid may be granted only if the recipient undertaking or group of undertakings is engaged in the implementation of a systematic and specific restructuring programme . . . capable of restoring its competitiveness and making it financially viable . . . under normal market conditions.[4]

A deadline for concluding the quota system was also extended first to 31 January 1984, then later to the end of 1985 in line with the policy on state aid.

The July 1983 proposals were based upon demand forecasts prepared earlier in that year.[5] These soon proved to be optimistic, however, leaving the Commission facing an uncomfortable situation — a fast-approaching deadline for the end of the Community steel regulations but no end in sight for the problem of over-capacity. In 1984, as revised forecasts were prepared, the Commission issued its first formal warning that 'capacity utilization rates would be unsatisfactory' if works closures were limited to the 26.7 m tonnes (from the 1980 capacity of 168 m tonnes) in conformity with the decisions which the Commission had taken in 1983 in application of the aid code.[6] The following year it considered forecasts to 1990 and called for a *further* 27 m tonnes of capacity to be closed by the end of the decade, in addition to the 26 m tonnes already closed (see Tables 2.15 and 2.16).[7]

Table 2.16 1985 European Community steel supply and demand forecasts for 1990

	Maximum production capacity required in 1990 (m tonnes)	*Surplus capacity from 1985 level (m tonnes)*	*%*
Crude steel	140.7	27.0	16.1
Hot rolled flat products	75.8	11.5	13.2
Long products	43.0	12.9	23.1
Cold rolled sheet	38.1	6.7	15.0
Coated sheet	16.6	1.8	9.8

Source: European Commission *General Objectives Steel 1990*

Note: Calculated on 85% utilization rate for crude steel, 80% for rolled products

In this context it was politically impossible unilaterally to abandon the quota system, although the Commission was keen to phase out some controls as evidence that its system of regulation was at least partially resolving the problem of over-capacity. At first sight paradoxically, the West German government, which had initially opposed introduction of the whole system, argued in favour of its retention. To remove regulation, it now maintained, would invite a price-war with state-subsidized competition pitched against the private West German steelmakers. Other countries were more concerned to secure an increase in their quota under the existing system. At the end of October 1985 agreement was therefore reached to carry part of the system over for a further year.

Under the terms of the extension the West German delegation gained at least the substance of what it wanted when subsidies were permitted to help plant closures in only two ways: either to cover up to 50 per cent of redundancy costs, or where companies actually left the sector altogether. The Commission secured its way by removing some products from the quota system: concrete reinforcing bars and coated sheet, accounting for 15 per cent of the output of products under control. The UK and, to an extent, France benefited from two flexibilities in the method of allocating quotas. A company could arrange a change in its quota for a particular product provided there would be no more than a 25 per cent increase in the amount of quota for that product, and the tonnage transferred did not exceed 10 per cent of the company's total quota. There would also be a 3 per cent reserve in the total steel quota for all companies, which could be allocated by the Commission either to companies with insufficient quota, or with a deteriorating relative position in the national market over two successive quarters.

During 1986 negotiations commenced over the structure of the steel quota regime from 1987 onwards. Initial Commission proposals published in September envisaged the abolition of quotas on another 20 per cent of the industry's output, reducing the proportion of products subject to control to 45 per cent. Four categories were identified where deregulation was possible: galvanized sheet, light sections, wire rod, and merchant bars. The Commission's desire to discontinue the whole quota system from the end of 1987 was also confirmed. In October the Commission endorsed these objectives with a warning that even further cutbacks might be required: 'There is still a structural imbalance between supply and demand and it might be necessary to introduce changes in the configuration of the industry going well beyond the envisaged capacity reductions'.[8]

This new proposal was unpopular with the big integrated steel producers and in November their association, Eurofer, put forward an alternative plan which entailed continuation of the existing quota regime until the middle of 1987 and replacement by an (unspecified) alternative system of controls from 1 July 1987 to 1990. During this period its members would close a further 11.9 m tonnes of surplus capacity. The draft plan called also for the creation of a social fund financed by the EC to compensate regions hit by steel closures. Quotas announced by the EC in December for the first quarter of 1987 excluded only galvanized sheet and light sections. Put another way, they still covered some 60 per cent of output. For most products quotas were set at very low levels, reflecting both continuing pressures of over-capacity and the seriousness with which the Commission was considering the alternative plan from Eurofer. In March 1987 further reductions were offered by the steel companies, amounting in total to 15.3 m tonnes of capacity. Crucially though, the plan offered no cuts in the production of hot-rolled coil, where over-capacity was greatest and national state assistance most apparent. The scheme was dismissed as inadequate by the Commission and the steel producers were given more time to offer still further cut-backs.

By May it was apparent that the gap between the cuts sought by the Commission and those offered by the steel companies could be narrowed no further, and the EC set in train the process of formulating new plans for the industry's future. Judged even by previous standards, this was to prove a long and tortuous affair. From the outset the Commission formally reaffirmed its desire to see production quotas abolished by the end of the year, although some flexibility was informally acknowledged for those products where over-capacity remained most pronounced. The inability of Eurofer producers to offer further cuts symbolized that the 'easy' closures had been made; further ones would depend upon politically sensitive decisions concerning the future of large, integrated sites. On the other hand the very real threat of ending production quotas was to give the Commission a powerful bargaining counter in negotiations with member state governments.

The Commission released its proposals in July. Under these, if adopted, quotas would remain in place until 1990 for flat products and heavy sections, but be abolished for wire rod and merchant bar at the end of 1988. A production levy was to be used to raise 600 million ECU (European Currency Unit) over three years to contribute to some of the social costs of restructuring and to provide an inducement to companies to close capacity. A further 870 million ECU aid package was envisaged from EC funds. The total cost of the scheme was compared to the estimated 5,000 million to 6,000 million ECU which the Commission anticipated would have to be spent in any event by member states and steel companies on closing the excess 30 m tonnes of capacity and shedding a further 80,000 jobs. When EC industry ministers met to discuss this proposal in September, the UK was the only member state fully committed *not* to accept continuation of quotas; but other elements of the package met with far less agreement. The ministers adjourned in disarray until December, appointing a three-man team of industry experts to prepare a report on the issue.

This was completed by mid-November. It concluded that the companies had failed to offer sufficient closures to justify continuation of the quota system. No fresh proposals had been made to break the deadlock. In response, the Commission prepared to abolish quotas on wire rod and merchant bar from the end of the year; on hot rolled coil from July 1988; and on heavy sections and heavy plate from the end of 1990, but only if companies produced what it saw as adequate guarantees of closures by mid-March 1988. If industry ministers could not agree on this or a compromise proposal, the Commission promised to end the entire quota system on all products with effect from 1 January 1988.

A protracted round of negotiations between the Commission and the Council followed and lasted throughout December. During these the Commission relaxed its stance on hot rolled coil, agreeing to prolong quotas for that product until the end of 1990 if producers guaranteed 7.5 m tonnes of capacity cuts, far more than the 5.8 m tonnes then on offer. This, and the enlarged total of 20.7 m tonnes of closures in all products now being offered (while still below what the Commission sought) proved sufficient to produce a temporary compromise. A formal timetable was at last agreed, for the abolition of quotas on wire rod and merchant bar from 31 December 1987; on hot rolled coil from July 1988 or later if fresh cuts were guaranteed; and on heavy plate and heavy sections from the end of 1990.

Some uncertainty remained, though. In return for its agreement on continuing quotas, the Commission required firm guarantees of closures by 10 June 1988. And this was ultimately to prove an insurmountable obstacle for steel companies. With the deadline expired, no closures had been offered in hot rolled coil, and cutbacks proposed in other sectors were well short of the Commission's demands. Karl-Heinz Narjes, EC Industry Commissioner, issued a call for the total abolition of the quota system:

> The conditions which we fixed for the extension of the quota regime have not been fulfilled and it will therefore expire. Steel firms should now stand on their own feet.
>
> (quoted in *Financial Times* 16 June 1988)

Later that month EC industry ministers agreed finally to abolish the quota system, with effect from 1 July. The shifting international coalitions which prolonged the crisis management system for so long had finally fallen apart.

More fundamentally, the Community steel quota regime had gradually proved inadequate as a mechanism for resolving the crisis in the industry. Apparently an embryonic supranational state was no more capable of managing a crisis of international dimensions in steel production than were the various national states in the Community. Taking the form that it did as the result of a collection of compromises, it is highly doubtful whether the Community's steel policy ever could have been sufficient in its own right to achieve such a task. This raises questions as to why the policy took the form that it did and as to the limits of Community steel policy. That it was devised and implemented at all is testament to the durability of the EC as an institution, the magnitude of the overcapacity, and the increasing evidence of national governments' inability to manage decline in a highly internationalized industry even through strongly interventionist policies such as nationalization. Community steel producers were further affected by the accession of new members as the scope of the crisis broadened across most grades of steel production.[9] It is interesting to speculate about what might have happened had there been *no* Community policies. It is also salutary to point out that crisis-management within the EC has acted on occasions to deflect criticism away from member state governments and expose many of the weaknesses within the EC itself. Partly in response, the Community became keen to disengage from intervention as soon as practically possible; but this proved to be even more politically contentious than the process of applying quotas in the first place. Clearly EC steel policy was unable to please all interested parties all of the time.

Other western Europe

The three major non-EC western European steel producers in 1985 were Spain (with an output of 14.2 m tonnes), Sweden (4.8 m tonnes), and Austria (4.7 m tonnes) (see Table 2.5). With its accession to the EC in 1986, Spain became firmly attached to the Community's policy framework. The two other major producers demonstrate strongly differing tendencies of production reorganization.

The way in which Sweden has rationalized its steel industry provides an exceptionally clear illustration of the opportunities and problems in this sector. Although total steel capacity is relatively small, at around 5 m tonnes annually, an unusually large proportion (around 30 per cent) is of special grades, making the country's evolving special steel producers among the major European competitors. Retrenchment and reorganization throughout the industry began well before most European steelmakers recognized the depth and severity of the crisis. The first major move, which cleared the way for rationalization in the special steels industry, came in 1978 when the largest ordinary steel producers merged to form Svengst Stal. The Swedish government took a half-share in this new company. By 1982 it had cut employment by a fifth to 14,000 and capacity by a quarter to 3.1 m tonnes. Losses of SKr 2.4 billion between 1978 and 1982 were turned into small but increasing profits thereafter.

Rationalization in special steels was concentrated in three product areas. By the mid-1970s the country's three leading tool steel producers had merged to form Uddeholm Tooling, the world's third biggest company in the sector after VEW of Austria and Thyssen of West Germany. In 1982 the high-speed steel producers joined forces as Klosters Speedsteel, a world leader with an annual output of some 15,000 tonnes and a 20 per cent share of the European market. And in 1984 the four stainless steel companies merged to form two units, eliminating overlapping production. Under a restructuring plan Avesta paid SKr 230 million each to Uddeholm and Fagersta to leave the sector, concentrating into one company European market shares of 40 per cent in stainless welded tubes and 30–35 per cent in stainless strip and hot rolled plate. Sandvik created a wholly owned subsidiary, Sandvik Steel, to specialize in production of seamless tubes, strip, and wire. The Swedish government agreed to provide some SKr 460 million in special financing to cover the agreements, mainly in the form of loan write-off, while the costs of laying off some 1,500 workers because of capacity closures were borne between Avesta and Sandvik. The restructured industry maintained a capacity of some 200,000–250,000 tonnes of stainless steel production a year, almost entirely for export to Europe. Finally, ball and roller bearing production was concentrated throughout this period at SKF, whose low alloy special steel division, a market leader in bearing steel, produced some 400,000 tonnes in 1982.

The stainless steel merger completed a reorganization of the industry into discrete product areas each controlled by major producers. So successful were the new ventures, however, that they were soon to fall foul of Sweden's position within the international system of trade. During 1983 the 1.2 m tonnes they produced included more than three-quarters of total Swedish steel exports (by value). With the proportion of special steels output to ordinary grades more than double the EC average, Sweden was clearly heavily dependent upon export markets. These countries' own producers were themselves rationalizing and seeking a degree of protection from competition. During 1984 Swedish producers were forced to agree to reduce exports of tool steel to the USA by some 20 per cent. Later the same year West German steelmakers made it clear that they too were considering a complaint, following a 27 per cent rise in imports of Swedish steel of all grades during 1983, to some 450,000 tonnes. And in October 1986 a group of US special steel companies filed anti-dumping petitions against Avesta and Sandvik — the fifth such action against Avesta in a ten-month period. The very success of Swedish steel producers' rationalization made them a target of international trade restrictions.

Partly in response the Swedish steel industry began to internationalize production. During 1984 negotiations took place between Avesta and China, with the latter seeking access to western technology and offering a guaranteed market for Avesta products. In early 1985 Uddeholm announced its intention to take a 20 per cent stake in a new, 15 million dollar special steels plant near Pittsburgh in the USA. Later in the same year SKF formed a new operational division for roller bearings in north America, with the intention of expanding in the US market from the production base there. And in 1986 SKF Steel merged with Ovako of Finland to form one of the biggest special steel producers in Europe with a workforce of 7,000, an annual production of 1.2 m tonnes, and a guaranteed market in bearing steels with the SKF parent company.

In many ways, then, Sweden illustrates several archetypical responses to economic change from a relatively small, advanced economy: early rationalization and investment in higher technology, dependence upon continued exports, and internationalization of production because of a threat to those export markets. By contrast the Austrian industry is dominated by the state-owned Voest Alpine, which is a bulk steel producer, also incorporating VEW and a broad range of engineering activities. It illustrates the failure of an alternative strategy, that of a substantial diversification out of steel. By 1985 its trading losses reached Sch 12 billion, attributed to a failed microchip venture, participation in the unsuccessful Bayou Steel Corporation in the USA, and calamitous results from a subsidiary involved in speculative oil deals. This unfortunate combination of circumstances led to the resignation of the chief executive and a new restructuring plan was hastily formulated, entailing 10,000 job losses at the parent company.

The USA

Steel producers in the USA have also evolved a range of strategies, with similarly varying degrees of success, in their attempts to break out of the vicious spiral of decline. The country is the third largest producer of steel after the USSR and Japan, and eight companies produced more than 2 m tonnes in 1985. The largest, US Steel, produced 15 m tonnes, second only to Nippon Steel of Japan (see Tables 2.5 and 2.6). The slump in domestic demand had particularly serious consequences, for it coincided with a dramatic surge in imports to the US market. In response steel companies have lobbied for trade protection, restructured their steel and non-steel operations alike, and focused increasingly on cost-cutting through labour contract renegotiation.

The susceptibility of the US steel market to imports has much to do with the early history and subsequent location of steel production capacity (see especially Markusen 1985: 73–100; 1986). From the later years of the nineteenth century, growth in steel was heavily concentrated in the north-eastern states, around Pittsburgh and Youngstown. After a short period of intense competition, a series of mergers resulted by 1900 in a near-monopolistic situation where one company, US Steel, controlled 65 per cent of output. At this time the industry also adopted a procedure for fixing prices known as Pittsburgh plus, or the basing point system. Prices quoted by all suppliers, whatever their location, were the Pittsburgh rate plus the cost from Pittsburgh to the market, effectively eliminating any advantage for a steel plant located nearer to the customer than Pittsburgh. This system of oligopolistic pricing greatly favoured the continued prominence of Pittsburgh as a production centre. In 1924 and again in 1938 government action forced the steel companies to designate additional basing points but the tendency towards spatial concentration remained. As late as 1945, over 90 per cent of all steel capacity was in the north-eastern states.

In 1948 the industry abandoned multiple basing point pricing, replacing it with a system whereby all firms automatically followed US Steel's lead in pricing. During this era therefore, companies were assured of a comfortable profit margin and faced little incentive to seek out new, more profitable locations; nor did they do so. Collusion between steel firms began to break down only towards the end of the 1950s as the USA's role in the world steel industry began to decline, and members of the United Steelworkers Union (USW) engaged in increasingly bitter action to secure some of the benefits of excess profits, culminating in a 116-day national steel strike in 1959. Belatedly a degree of decentralization occurred with the dispersed development of smaller, electric arc steelworks, but on the eve of recession this trend had had only a marginal impact on what remained a geographically imbalanced locational pattern.

As a result of oligopolistic control of prices, then, the US steel industry entered the recession both peculiarly vulnerable to imports to serve the west coast and southern states' manufacturing capacity, and to closures within the increasingly antiquated north-eastern heartland. The early history of the industry therefore impacted on its later fortunes in a particularly concentrated phase of restructuring, aimed at internal reorganization and diversification in the face of an import threat made particularly acute by the degree of global over-capacity.

Just as in the European Community, the US steel industry saw a dramatic slump in output in 1975, down from the 109 m tonnes of the previous year to 80 m tonnes, followed by a steady climb back up to 100 m tonnes in 1979 and two further slumps to 84 m tonnes in 1980 and to 62 m tonnes in 1982 (see Table 2.17). Similarly, after a decade of crisis, output remained well below that achieved in 1974, in this case more than 30 per cent down. The first year of overall industry financial deficit, 1977, was followed by a four-year period which saw 12.5 m tonnes of capacity closed down. But after 1981 the industry really felt the full brunt of domestic collapse and an import surge, and the pace of restructuring accelerated dramatically (see Clark 1987a).

The largest company, US Steel, embarked upon a programme of closures which cut capacity in its most vulnerable areas, particularly long products such as bar, wire, and rails, and concentrated

Table 2.17 US steel industry: domestic shipments and imports, 1974–84

	Domestic shipments (m tonnes)	*Imports (m tonnes)*	*Import market share (%)*
1974	109	16	13
1975	80	12	14
1976	89	14	14
1977	91	19	18
1978	98	21	18
1979	100	18	15
1980	84	15	16
1981	88	20	19
1982	62	17	22
1983	68	17	21
1984	73	26	26

Source: Financial Times 26 July 1985

resources upon flat rolled products. In line with this market strategy it briefly considered for a few months in 1984 a merger with National Intergroup, a smaller producer heavily biased towards sheet steel for the Detroit car industry. When the US Justice Department's anti-trust division made clear its opposition to a similar plan from LTV and Republic Steel on the grounds of undue market concentration, US Steel and National Intergroup abandoned their proposals. At the same time US Steel adopted a policy of diversification out of steel and into oil and gas, including the purchase of Marathon Oil in the early 1980s and of Texas Oil and Gas in 1986. The shift of direction was exemplified in the company's change of name to USX later in that year, symbolically dropping the 'Steel'. By then more than half the group's sales came from its oil and gas business.

LTV occupied second place amongst US steel producers in 1985. It was a company which had consistently grown through merger, most recently in 1984 with the acquisition of Republic Steel, after the terms were altered under pressure from the Justice Department to exclude certain product areas. It too diversified out of steel, into aerospace and industries linked to defence. By 1986, however, the company faced an acute crisis, partly due to problems arising from the acquisition of Republic Steel's debts, and it was forced to file for protection under Chapter 11 of the US Bankruptcy Code. This gave the company temporary protection from creditors while it worked out a reorganization plan.

The third biggest US steel company in 1985, Bethlehem Steel, stuck to the sector and invested heavily but cut its workforce equally dramatically. Like Inland and Armco, the next largest companies, over-capacity was not so much a problem as the variable costs of production, and especially labour productivity. National Steel, the next biggest company, took this concern to raise labour productivity to unprecedented levels. In 1984 it sold off its Weirton Steel plant under an Employee Stock Ownership Plan, securing an 18 per cent pay reduction in the process. In the same year 50 per cent of National Steel, the wholly owned subsidiary of a newly created holding company, National Intergroup, was sold to the Japanese steel group Nippon Kokan. This brought an infusion of capital, technology, and management which was to prove particularly transformative for the company's collective bargaining arrangements.

Wheeling-Pittsburgh, one of the most vulnerable US steel companies (despite a technology deal with Nisshin Steel which gave the Japanese company a 10 per cent stake) also filed for protection under Chapter 11, in 1985. Although one of the smaller companies, a proposal to cut wages from 21.40 to 17.50 dollars an hour in order to restore profits soon thrust it into the limelight. For by this time all other US companies had come increasingly to focus on enhancing labour productivity and cutting wages as a means of raising profitability. The Wheeling-Pittsburgh proposals were regarded with considerable interest, therefore, not least by the steelworkers' union. In July 8,200 workers walked out on strike, and stayed out for three months. They finally returned on conditions which included a wage cut to 18 dollars an hour.

The Wheeling-Pittsburgh dispute and the terms of its conclusion were symptomatic of a growing concern on the part of US steel companies to cut wages and benefits payments. In part the trend had been set at Weirton Steel, but this deal had been negotiated by the local Independent Steelworkers' Union, known by the USW for its close relations with local management. With wage cuts forced through at Wheeling-Pittsburgh, though, the USW was on the defensive. Most other major companies were already formulating plans for a new era and style of collective bargaining. In March 1985 Bethlehem Steel had signed a new equity and profit-sharing contract with the USW at its Johnstown Plant in Pennsylvania. The agreement included a 3 dollars an hour wages and benefits cut, to 20 dollars an hour; a 25 per cent drop in the workforce to 2,200; and the introduction of greater working flexibility on the shopfloor. Just as significantly, the deal was concluded outside the conventional company-wide bargaining arrangements, giving the plant a highly decentralized industrial relations structure for the first time.

These moves were important, for since the late 1950s collective bargaining in the steel industry had been the preserve of a co-ordinating committee set up by the twelve largest companies, to negotiate and pay the same wages and benefits throughout all their plants. Under this system, the USW had negotiated a three-year contract with one of the leading employers which then became a model for the rest of the industry. In May 1985, however, in formal recognition of the emerging situation, the companies dissolved the committee, throwing into question prospects for the end of the then current three-year deal in 1986.

In response to this uncertainty, the USW commissioned a key report, 'Confronting the Crisis', which analysed the prospects for the US steel industry. This came to an unusual and unambiguous conclusion, for a trade-union commissioned document: that the industry faced real financial difficulties. Even more exceptionally, it recommended the union's local negotiators to fight for the preservation of steelmaking through co-operative agreements, including if necessary cash concessions.

This (at first sight unorthodox) model was to be widely followed. In April 1986 workers at LTV agreed to a cut in wages and benefits from 25 to 21.50 dollars an hour, in return for a profit-sharing deal. The following month Bethlehem Steel proposed company-wide average wage and benefits cuts of 2 dollars an hour, to 22.50 dollars an hour, again in return for a profit-sharing agreement. In June workers at Inland Steel reached agreement on a three-year wage freeze and a 0.50 dollars an hour benefit cut, lowering total labour costs to 21.50 dollars an hour.

The greatest shift in collective bargaining arrangements, though, was to take place at National Steel, the company which had been partly sold to the Japanese group Nippon Kokan in 1984. By 1986, while other companies were signing deals with wage and benefit cuts, and other changes designed to improve labour productivity, National Steel completed negotiations with the USW over its innovative Co-operative Partnership (see Clark 1987b). This was a far more wide-ranging agreement which encompassed collective commitment to improved labour productivity, wage and benefit cuts of 1 dollar an hour, flexible working, and reassignment across job classifications. In return the company promised new investment and limited job guarantees. The deal provided for a Productivity Gain Sharing Plan, offering bonuses on the basis of labour productivity improvements, and a profit-sharing scheme. It also introduced joint management-labour problem-solving committees, effectively decentralizing control of day-to-day operations. Overall it tied employees to the efficient functioning of the company in a clear imitation of the Japanese model of industrial relations.

In the mean time talks at USX remained deadlocked, and in August 1986 the company faced a complete stoppage as the USW refused to cede the company's demand of a 3.30 dollars an hour wages and benefits cut to 25.20 dollars an hour. The strike did not end until January 1987, after a long and bitter six months, when the USW accepted a new four-year labour contract with a wages and benefits cut of 2.45 dollars an hour in the first year. The following month USX made plain its determination to make sharp cuts in steelmaking operations, with an announcement of plans to axe capacity from 26 to 19 m tonnes, close five plants indefinitely and shed 20 per cent of its steel workforce.

Clearly, despite an intense period of rationalization, with its many manifestations of closure, merger, diversification and new patterns of collective bargaining, the US steel industry remained in crisis. In such circumstances, attention was increasingly drawn to the country's trade deficit in steel products, most especially by aggrieved domestic producers. Yet in fact the US trade gap represented only one feature, albeit a significant one, of a major new pattern which had been emerging in world steel trade for some time.

Trade in steel

New trends in international steel trade

From the mid-1970s, as steel production continued to expand in the NICs and faltered or declined in the more advanced economies, patterns of international trade which had been evident in the 1960s took on a new and growing significance. Within the complex overall flow of imports and exports, several general trends could clearly be discerned (see also OECD 1985). The first and most significant of these was an almost continuous increase in the proportion of steel output traded internationally. From 1960 to 1974 exports grew annually at an average rate of 10 per cent while world output grew at an average annual rate of 6 per cent; from 1974 to 1983 exports grew by 2 per cent annually while output stagnated. By the start of the 1980s, there had also been a substantial redirection of trade patterns, with an expansion in NIC exports (especially to other NICs) and a partial diversion of OECD exports to the NIC markets in response to growing protectionism within the developed countries (see Table 2.18).

In 1983 the world's ten largest exporting countries included not only Japan and five of the European Community's founder-members, but also South Korea and Brazil from within the ranks of the NICs (Table 2.19). The top ten importing countries included

Table 2.18 International steel trade patterns, 1960–83

World exports: three-year averages	*1960–62*	*1981–83*
% from		
OECD	90	77
Non-OECD market economies	1	15
Centrally planned economies	9	8
% to		
OECD	49	41
Non-OECD market economies	39	43
Centrally planned economies	12	14
Total (m tonnes)	25.6	102.7

Source: OECD

Table 2.19 Major steel exporting and/or importing countries, 1983 (m tonnes; top ten countries)

Exports		*Imports*	
Japan	30.9	USA	15.2
West Germany	15.7	West Germany	11.5
Belgium/Luxembourg	11.4	China	9.6
France	9.5	USSR	9.1
Italy	7.4	France	7.6
USSR	5.8	Saudi Arabia	5.8
South Korea	5.7	GDR	4.8
Spain	5.7	Italy	4.8
Brazil	5.2	UK	4.1
Netherlands	4.2	Belgium/Luxembourg	3.3

Source: International Iron and Steel Institute

four of the major European Community exporters, demonstrating the relative openness of their markets (within EC boundaries at least). NIC importers included China and Saudi Arabia, but easily the most significant single importer was the USA. Its massive steel trade deficit served only to underline the political significance of a series of import restrictions in the USA, most especially on trade with the European Community.

United States–European Community steel trade agreements and their implications

There have been three major periods of crisis in the recent history of United States–European Community trade agreements: 1968, 1977, and 1982–84 (see also Jones 1979; 1985; 1986; Mueller and Van der Ven 1982; Walter 1979; 1983). Protectionist sentiment arose on each occasion as a result of surges in exports from a small number of easily identifiable countries. The US government formulated responses to mollify domestic producers, yet attempted to avoid generating similar retaliatory measures from other governments. But once such policy instruments were put in place, exports which previously had been destined for the US were switched to other markets, thereby generating a similar crisis elsewhere — part of a 'protectionist spiral' in steel trade (Jones 1985).

The first round of steel trade restrictions was negotiated in the wake of a slump in US domestic demand in 1967. This had the effect of making rising import penetration more readily apparent. Both the steelworkers' union and the steel companies lobbied Congress, proposing comprehensive five-year bilateral import quotas on all supplying countries. US Steel also filed a complaint against EC countries alleging that their exports were unfairly subsidized. The US government was caught in a dilemma, anxious to appease the still powerful steel interests but mindful of the long history of oligopolistic pricing practices of US steel companies. It was saved by an intervention from the Japanese Iron and Steel Exporting Association, fearful of the potential effects of a bilateral restrictive deal. The Japanese exporters proposed instead Voluntary Export Restraint (VER), and EC producers quickly agreed this was the best arrangement possible in the circumstances. Under it, exporters were allowed at least some role in setting their own quotas. The first formal agreement ran from 1969 to 1971, limiting Japanese and EC exports to the US to 5.75 m tonnes annually from each party.

In 1970 prices rose uniformly in the USA, the EC, and Japan, and US imports did not increase. The following year though, prices fell in the EC and Japan while those in the USA rose, leading to a further surge in imports. A second round of negotiations began, leading to a new VER agreement for the period 1972–74, which set new limits of 7.3 m tonnes annually from the original EC members plus the UK, and 5.9 m tonnes annually from Japan. For the first time also the deal included specific (but still voluntary) limits on stainless, alloy, and high-speed grades of steel. It was generally a more stringent set of controls, but was never really put to the test since devaluation of the dollar in 1971 and 1973 caused all import prices to the USA to rise; and high domestic demand in Japan and the EC reduced the potential surplus to be exported elsewhere.

At the same time the whole basis of the Voluntary Export Restraint agreement was subjected to a lengthy legal challenge within the USA, on the grounds that it violated both domestic anti-trust legislation and the US Constitutional commitment to free trade. The first aspect of the case was dropped at an early stage and the second ultimately failed, but the action raised a number of significant legal and political questions. As steel demand boomed in 1974, the US government quietly dropped the whole VER idea.

Such profitable market conditions proved to be short-lived and several developments led the way to a second crisis in US steel trade policy. As global over-capacity became increasingly apparent, producers in Japan and the EC turned once again to the US export market. Japanese exports were particularly concentrated on the USA, after a deal was concluded limiting Japanese exports to the EC. Just as in 1967, the US steel companies and the steelworkers' union advocated some form of protection; and in 1977 US Steel filed an anti-dumping complaint, this time against the Japanese exporters. Along with the EC companies, Japan proposed in turn a new VER agreement; but this proved unpopular in the USA in the light of the record of the previous voluntary agreements. The protectionist lobby eventually gained its way, with a new system, the Trigger Price Mechanism (TPM), to be implemented from March 1978. This induced 'voluntary' restraint from exporters by establishing a system of dumping reference prices, sales below which would lead to an investigation by the US government, with the threat of further sanctions to follow. It was a policy more powerful than the old system of VERs, yet still within US anti-trust legislation.

The TPM system also provoked a fresh reaction within the EC. The 1977 Davignon Plan had set minimum prices on domestic production and taken the first steps towards import regulation. By the end of that year a Basic Price System was introduced on imports. Similar to the TPM, it established 'fair value' reference prices below which an investigation would be implemented. Unlike the US scheme though, the EC proposals included an offer to withdraw the import pricing system if exporters were willing to negotiate voluntary restraint agreements based on a given tonnage. By the end of 1978 a large number of these deals had been concluded.

The TPM was introduced into an increasingly difficult market environment, and proved gradually to be inadequate to the task set for it. In 1980 US Steel filed an anti-dumping complaint against several EC producers; since this breached the terms of the agreement between US steel companies and the US government, the latter immediately suspended the TPM. It was temporarily reinstated later that year when US Steel withdrew the suit, but proved increasingly unpopular throughout 1981. The system was finally killed off by a major export surge from the EC. In 1982, as US steel output slumped to below one-half of capacity, fresh anti-dumping suits were filed and the TPM suspended indefinitely. A major political row ensued between the USA and the EC, resolved only towards the end of that year.

Under this new agreement the EC agreed to limit exports to the USA in ten specific product categories, in exchange for a withdrawal of forty-five charges of dumping levied by eight US steel companies. It was initially intended to run until December 1985. Almost immediately, though, the EC announced new limits on imports into the Community, further continuing the protectionist cycle. Four US companies argued that the EC's fresh move would divert other countries' exports to the USA; they therefore pressed for fresh import controls with countries such as Japan, Taiwan, Brazil, and South Korea. In the wake of huge steel company losses, a massive protectionist lobby developed, culminating in the announcement of new measures in 1984. These included a limit on imports from all sources set at 18.5 per cent of the domestic steel product market, with a further allowance of 1.7 m tonnes of semi-finished steel annually. This was to be achieved through five-year agreements on a product and country basis. Most significantly, with the installation of a new licensing system to monitor imports, the process of restraint could no longer be labelled 'voluntary'. The barrier was further strengthened the following year when new products, especially special steel grades, were added to the list of those facing restraint.[10]

The cycle of protectionism had therefore taken another turn, with a fresh import surge generating government trade restriction, retaliatory measures and export displacement, and leading to further, more stringent controls. The significance of the EC–USA deals has been far-reaching, not just for other steel trade agreements but also in stimulating and fostering co-operation between US steel companies and the steel unions. Equally the growing importance of NIC exports to the USA epitomizes the changing geographical balance of the global steel industry and its increasing interconnectedness. One of the cruellest ironies of this new threat to the stability of previous trade patterns lies in the role of international lending agencies, which have imposed steel export strategies on countries like Brazil. In this way the question of international steel trade becomes one part of the broader mosaic of the international development process (see also Markusen 1986: 456). This is also apparent from an examination of the patterns of indirect trade in steel, in the form, for example, of motor vehicles.

Indirect trade in steel and the competition from other products

Indirect trade in steel is of such significance for the world's steel producers that the International Iron and Steel Institute produces a triannual report on the subject (see IISI 1985). This is based on a sample of fourteen major countries which account for over 90 per cent of the western world's engineering exports by value, and 78 per cent of its steel output by volume. In 1982 these countries produced 261.7 m tonnes of steel, of which 94.9 m tonnes was exported directly and a further 56.6 m tonnes indirectly. The motor vehicle industry played a dominant role in this indirect steel trade. Cars, commercial vehicles, and components amounted to 23.4 m tonnes, or 41 per cent of the total. In many respects motor vehicle production is an international operation, dominated by a few major multinational corporations able to source components and assemble vehicles throughout the globe (see Hudson and Sadler 1987a). The motor vehicle industry is significant in another sense too, for it not only helps to shape new geographical patterns of steel production, but is also illustrative of the growing importance of alternative products to steel (see also IISI 1983).

In the mid- to late 1970s the pressure on vehicle manufacturers to increase fuel efficiency was so great that substitution of lighter materials for steel was considered by designers even if this meant an increase in overall production costs. With fuel efficiency less of an urgent issue, cost-effectiveness is now a prerequisite for material substitution. The greatest physical advantage of steel remains its strength and rigidity for weight. Aluminium is light and corrosion-resistant and in certain alloy forms has good strength for weight ratios, but is expensive and difficult to weld. Its main use has been in the casting of complex parts where it has substituted for iron and non-ferrous metals rather than steel. Plastics afford greater design freedom, are corrosion-resistant, and are often easy to form. However, in their unreinforced state they have low strength and poor heat resistance, which is a handicap in the engine area. In their reinforced form they have found some body applications.

A more recent innovation is the increasing use of a broad range of fibres, often made not of glass or carbon but of ceramic materials such as boron, silicon nitride, silicon carbide, and oxides of aluminium, zirconium, and silicon. A broad range of applications is envisaged for these new composites in industries such as aerospace, motor vehicles, and specialized machinery. They offer better physical properties such as strength or heat resistance, although some industry analysts caution that the high costs derived from expensive raw materials and labour-intensive fabrication will represent a serious barrier to their use in all but the most advanced aerospace market. Boron fibre, for example, is some 500 times more expensive than steel, and ten times dearer even than carbon fibre.

Yet new materials are finding their way into motor vehicles. In 1984 GKN announced the successful completion of tests on a glass fibre and epoxy resin composite (having discarded carbon fibre) as an alternative to the production of conventional steel springs used in commercial vehicles. The new material produced 50 per cent weight savings and GKN embarked on a £6.4 million investment programme to install a plant at Sankey near Telford in the UK, capable of making 500,000 springs a year (for a world market of some 20 million commercial vehicle springs annually). Production was so automated that only 100 jobs were created. In 1986 GKN also announced that it was to develop and produce plastic composite suspension systems for cars, up to 70 per cent lighter than conventional systems, to be commercially available within five years.

In their chief market area, then (and in other sectors — see for example NEDO 1985), steel producers face a major and often under-estimated challenge, both from the internationalization of production and from the competitive threat of alternative materials and products. Yet the steel industry is integrated with the rest of the economy not only through its market, but also through the supply of raw materials. We turn now to examine how changing patterns of steel production and consumption have been reflected in the major raw material supply sectors of iron ore, coking coal, scrap, and alloy materials.

Changing patterns of supply in the raw material sectors

The pattern of supply in the iron ore sector has undergone a number of changes since 1974 (see also Bradbury 1982). As global steel production stagnated, total iron ore output declined, with the increased use of higher grade ore and of electric arc steel furnaces (see Table 2.20). The collapse of the steel industry in the advanced economies has also been mirrored by a decline in the traditional sources of supply of iron ore. The world scene was dominated throughout the period by the USSR, with its output strongly geared to domestic consumption. Brazil, Australia, and China were other major producers. By contrast there has been spectacular decline since 1974 in the USA, France, and Sweden.

A similar picture is apparent from the world's exports of iron ore (see Table 2.21). The dominance of two countries, Australia and Brazil, is even more striking. Together, they accounted for roughly one-half of iron ore exports in 1983. Yet they have followed different paths since 1974. Brazil has expanded while Australian exports have declined (along with those of Canada, Sweden, and France). The reasons for the divergent paths of these two major exporters are revealing.

The Australian iron ore mining industry originally grew on the strength of demand from Japanese steel companies, formalized through a series of contractual commitments in the 1960s. This market is particularly significant, accounting for around one-third of the world's iron ore imports. The recent decline in Australian output is due to a collapse in demand from this source, and to the availability of alternative, newer sources of supply — which has in turn been encouraged by the Japanese (see Table 2.22). Chief amongst these is Brazil. The state-owned mining company, CVRD, saw its major new site at Carajas (in the south-east Amazonian

Table 2.20 Major iron ore producers, 1974–83, m tonnes (rank)

	1974	*1983*
USSR	224.9 (1)	245.2 (1)
Brazil	80.0 (4)	92.1 (2)
Australia	97.0 (2)	79.0 (3)
China	60.0 (5)	70.0 (4)
India	35.5 (9)	47.6 (5)
USA	85.4 (3)	38.6 (6)
Canada	47.3 (7)	30.0 (7)
South Africa	11.6 (13)	16.6 (8)
France	54.3 (6)	16.0 (9)
Liberia	25.0 (11)	15.4 (10)
Sweden	37.0 (8)	13.5 (11)
Venezuela	26.7 (10)	9.7 (12)
Mauritania	11.7 (12)	6.6 (16)
Chile	10.3 (14)	5.2 (17)
Total	898.4	746.1

Source: International Iron and Steel Institute

Note: Includes all countries producing in excess of 10 m tonnes in either 1974 or 1983. 1983 rankings completed by North Korea 8.5 m tonnes (13), Spain 7.8 m tonnes (14), and Mexico 7.5 m tonnes (15).

Table 2.21 Major iron ore exporters, 1974–83, m tonnes (rank)

	1974	*1983*
Australia	83.7 (1)	74.3 (1)
Brazil	59.4 (2)	70.0 (2)
USSR	43.3 (3)	34.0 (3)
Canada	37.4 (4)	25.5 (4)
India	22.2 (8)	20.7 (5)
Liberia	25.7 (7)	15.4 (6)
Sweden	33.1 (5)	14.3 (7)
South Africa	2.9 (15)	7.8 (8)
Venezuela	26.3 (6)	7.6 (9)
Mauritania	11.7 (10)	7.4 (10)
Chile	9.4 (12)	4.7 (11)
France	19.8 (9)	4.6 (12)
USA	2.4 (16)	3.8 (13)
Total	407.4	303.1

Source: International Iron and Steel Institute

Note: Includes all countries listed in the main body of Table 2.20 except China (zero exports). 1974 rankings completed by Peru 9.6 m tonnes (11), Angola 6.4 m tonnes (13), and Algeria 3.2 m tonnes (14).

Table 2.22 Japanese steel industry's raw material imports, 1960–80 (m tonnes)

Iron ore	*Total*	*From:*	*Australia*	*India*	*Chile and Peru*	*Brazil*
1962	22.1		—	20.3	24.0	2.2
1965	38.8		0.5	19.9	28.0	2.0
1970	102.0		35.9	16.1	15.5	6.7
1975	131.7		48.1	12.8	8.0	17.8
1979	130.3		42.4	13.1	7.4	20.1

Coking coal	*Total*	*From:*	*Australia*	*USA*
1960	5.7		ND	ND
1965	14.3		5.9	6.3
1970	44.4		14.2	23.7
1975	57.7		22.4	22.2
1980	58.0		24.0	18.5

Source: Hogan 1983

basin) come into operation in 1986. For an investment of 3.6 billion dollars, this has the potential to produce 35 m tonnes annually of high grade, easily mined ore, shipped out via a new 500 km rail link from a new port at São Luis, capable of accommodating ships of up to 280,000 tonnes capacity. With total reserves estimated at 18,000 m tonnes, a potential output equal to one-tenth of total annual world exports and low operating costs, this single site alone is potentially capable of exerting a strong influence on the world price of iron ore.

The opening of this and other alternative sources of supply, along with declining demand, has led to dramatic contraction amongst several of the longer established producers. For example, the output of the Swedish state-owned company LKAB halved from 27 m tonnes in 1979 to 13 m tonnes in 1982 as a consequence of the collapse in the European Community steel industry. The Canadian industry too has suffered, despite the negotiation of a number of long-term contracts in the early 1970s. In 1975, for instance, the British Steel Corporation joined with Sidbec, the Quebec government-owned steel company, to develop a new iron ore mine, Sidbec-Normines, at Fire Lake. The intention was to secure access to this key raw material at a time of potential shortage. Yet the mine began production in 1979 when world demand, and prices, had slumped. Under the terms of the deal BSC was obliged to take some of its output at fixed prices, well above the newly prevailing world market levels. The 2.5 m tonnes supplied annually from Fire Lake cost BSC an extra £20 million each year. In 1984 therefore it renegotiated the agreement and at a cost of £103 million closed the mine completely. Not only had it been an expensive miscalculation, but also it graphically demonstrated the changing balance between supply and demand in the iron ore industry.

Historically the iron ore industry has been characterized by a high degree of control from the steel companies, either directly through ownership linkages, or indirectly through purchasing power (the former dominant at times of expansion, the latter in periods of decline). Prices are negotiated annually between the major mines and the largest steel companies. In the early 1980s growing over-supply led to a dramatic price fall, reaching 20 per cent over 1983–84. By 1987 exceptionally sensitive negotiations led to five leading iron ore companies — CVRD and MBR from Brazil, BHP Iron Ore and Hamersley Iron of Australia, and Sweden's LKAB — warning the Japanese steel producers not to press their claim for a 10 per cent price cut. If implemented it would, they argued, lead to a 'fundamental destabilisation of the world iron ore industry' (quoted in *Financial Times* 19 February 1987). The urgency of their response, and the final terms of settlement — a 5 per cent cut — demonstrated both the growing fragility of the iron ore supply industry and the continuing supremacy of purchaser over raw material producer in an over-supplied market.

The iron ore sector, then, has been characterized by a growing shift from the advanced world to the NICs, and by attempts from the steel companies to maintain either ownership or market power. The Japanese steel industry in particular has acted to encourage diversity and security of supply, first from Australia, later from Brazil. By the mid-1980s market conditions indicated that prices would at best stabilize around depressed levels, while the process of changing geographical balance was likely to be a continuing one.

A similar pattern is evident in the steel industry's other bulk raw material sector, coking coal. New reserves were opened up in the 1960s and 1970s, especially in Australia, largely in response to an extremely optimistic set of forecasts from the Japanese steel companies. By 1980 Australian mines supplied 24 m tonnes of coking coal to Japan's steel mills (see Table 2.22). Around this time the Japanese companies sought to diversify their sources of supply, fearing the potential effect of over-dependence upon one raw material producer. This led in 1981 to the signing of terms for a new coking coal mine, Quintette, in Canada. Japanese steel companies backed the project financially and agreed a fifteen-year supply contract at prices above prevailing world market levels. The mine opened in 1984. Yet just as in iron ore, growing global overcapacity in the 1980s led to price falls, mine closures, and job losses in the coking coal industry. The Quintette project was re-evaluated and with its future in jeopardy, price cuts were enforced. Others were not so lucky; demonstrating once again the instability of the raw material sectors, especially at a time of recession in steel production.

Modern steel production technologies do not, however, rely just on natural raw materials but wholly (in the case of electric arc furnaces) or partly (in the case of basic oxygen converters) on steel scrap inputs as raw material. Patterns of trade in the scrap industry arise from a complex balance between production (including scrap arising in the steel production process itself) and consumption. One illustration of the relationship between scrap production and trade, and the steel industry more generally, is from the UK. In late 1979 the UK government lifted (but did not abolish) an export quota and licensing system which had governed scrap exports for nearly twenty years. During the late 1970s more than 90 per cent of scrap handled by British merchants was sold inside the country. Since that time the rise of the scrap merchant as exporter almost exactly paralleled the decline of the UK steel industry and the rise of foreign producers, especially in Spain. Between 1979 and 1982 the amount of Spanish steel produced in electric arc furnaces rose by 20 per cent to 6.7 m tonnes. The producers needed more scrap, but the two traditional European exporters, West Germany and France, were tied to Italian plants in the Brescia region. Britain's big four merchants, 600 Group, Bird Group, Coopers (Metals), and Mayer Newman, filled the gap. They spent £10 million equipping deep-water berths at Tilbury, Cardiff, and Liverpool with scrap-handling equipment and in 1983, for the first time, UK exports of scrap (at 3.8 m tonnes) exceeded home sales of 2.9 m tonnes. Over 3 m tonnes of these exports were to Spanish steel producers.

The alloy metals required by the steel industry are in a very different category. They are available from very few sources and are of great strategic significance. They include chrome, which is vital for stainless steels; manganese, used in refining bulk steels; and vanadium, incorporated originally in tool steels but more significantly now in high-strength low-alloy steels (see IISI 1983). In 1978–79 South Africa had an 18 per cent share of world manganese ouput, 26 per cent of chrome production, and 35 per cent of vanadium output, giving it a strong global position. Yet as Bush *et al.* (1983) argue, the dependence of many western countries upon South Africa is by no means necessary or complete. It is as much a function of the activities of multinational mining corporations as of the lack of adequate reserves elsewhere. There is, therefore, a complex web of economic and political power surrounding production of and trade in these minerals.

The key raw material industries, then, have displayed a range of trends alongside the steel industry's dramatic transition. In the bulk sectors of iron ore and coking coal, new producers have emerged, with the decline of the old mirroring the new patterns of the consuming industry; while over-capacity since the mid-1970s has served to emphasize the power of the consumer in a buyer's market. The scrap metal industry has fluctuated in its production and trade, and the supply of key alloy materials has been concentrated on one source of supply, South Africa, as a result of multinational mining company activity in response to the lure of cheap labour. An over-riding impression though, is of the turmoil engendered by first the runaway expansion, then the traumatic collapse, in the consuming sectors of steel production; itself in chaos as a result of changes in its own market areas and sectors. The geography of economic restructuring, through its inter-sectoral linkages, is wide-ranging and uneven in its impacts.

Summary

In this chapter we have described the main patterns of production in the global steel industry. We have focused upon the contrasting fortunes of the older industrial countries and the newly industrializing economies and considered how international trade has come to play an increasing role in the evolving geography of steel production. We have examined how indirect trade in steel (steel traded in the form of manufactured goods) is also related to the question of how much steel is produced, where, and by whom. On the other side of the production equation, we have analysed changing trends of supply to the steel industry of key raw materials, especially coking coal and iron ore. One consequence of these developments has been a dramatic reduction in capacity and employment in steel production in many, if not most, advanced economies. In the following chapter we go on to consider the implications of this for steel producers and steelworkers in these countries, by describing and analysing in detail some of the ways in which proposals to close steelworks and shed thousands of jobs in localities built around the steel industry have been contested.

Notes

1 We have not considered a number of other countries of considerable if secondary importance. They have been described elsewhere: on Australia, for example, see Donaldson 1981; Donaldson and Donaldson 1983; on Canada, see Bradbury 1985; on the USSR, see Zum Brunnen and Osleeb 1986.

2 The Commission's role was described as follows by François-Xavier Ortoli, Vice President of the ECSC Commission, in 1976:

> The funds made available to undertakings derive to a very large extent from the United States market and the capital markets which offer the most advantageous terms. The reactions encountered on the capital markets, particularly that of the United States, have confirmed the Institution's credit-worthiness. . . . Most beneficiary undertakings have thus obtained borrowed funds on terms, with regard to interest rates, maturities and amounts, which it would probably have been very difficult or impossible for them to find for themselves on the open financial market.
>
> (ECSC, *Annual Financial Report* 1976: 8)

3 Bull. CEC, 6-1981, 1.4.1–1.4.3. An initial Commission decision on rules for state aid to steel producers had been prepared on 1 February 1980 and was due to expire on 31 December 1981. This was replaced by a Commission decision published on 7 August 1981 which recognized five categories of aid:

(a) investment
(b) closure
(c) continued operation, not permitted after 31 December 1984
(d) research and development
(e) emergency, not permitted after 31 December 1981

For details see Dominick 1984, especially pp. 392–401.

4 Bull. CEC. 6-1983, 1.1.1–1.1.4.

5 *General Objectives Steel 1985* Com (83) 229, 22 April 1983.

6 *Comments on the General Objectives Steel 1985* Com (84) 89, 20 February 1984, p. 1.

7 *General Objectives Steel 1990* Com (85) 450, 31 July 1985.

8 *Report on the General Objectives Steel 1990* Com (86) 515, 7 October 1986.

9 By far the most significant steel producer amongst these was Spain, which produced 14.2 m tonnes in 1985 against just 1.0 m tonnes in Greece and 0.7 m tonnes in Portugal. In that year the Spanish government evaluated its plans for re-shaping the steel industry, due to be completed by 1988, in the light of Community objectives. It made it clear that as the country had a relatively low consumption of steel per head, the industry would continue to export its products aggressively

both within and outside the Community, and protect the home market with import quotas on certain products, even from other EC member states. Throughout the transitional period of Spanish accession, overcapacity in its steel industry would affect the EC steel market. (*General Objectives Steel 1990: position of the Spanish and Portuguese authorities* Com (85) 774, 18 December 1985.)

10 The addition of special steel was of significance because many producers had sought to increase the *value* of their sales within the USA through increasing the *proportion* of their exports of special grades of steel without increasing the *total tonnage*; thereby producing a deepening crisis for the US special steel producers (for details see Hudson and Sadler 1987b).

Chapter three

Contesting steel closures: twists and turns on the path of decline in western Europe and the USA

Introduction

While the process of creating new patterns of production and trade in the steel industry was an international one, it also had significant local and regional dimensions. With global demand static, growth in the newly industrializing countries (NICs) was matched by decline in many former steel centres. The pressures of competition increasingly led to question marks being placed over the future of whole steelworks and the communities which had historically grown up around them, especially during a particularly concentrated phase of restructuring from the late 1970s onwards (see Hudson and Sadler 1986).

Since the 1950s the steel industries of western Europe and north America had been undergoing locational change in response to the new market conditions offered by imported high-grade iron ore and coking coal, as opposed to the historically significant lower-grade indigenous raw materials. With technological changes also favouring the construction of a new generation of larger works, investment became concentrated at fewer, coastal locations (see Fleming 1967; Warren 1967). In a context of economic boom and growing demand for steel, such developments could easily be accommodated in this period via either overall capacity expansion or selective transfer of workers from older, inland facilities to the newer, often green-field plants.

The impact of the post-1973 recession was felt heavily, though, in the steel industries of the advanced economies, and accentuated by the extent of continuing NIC producer growth. National governments were increasingly drawn in to the affairs of many steel companies via financial support or outright nationalization. In the process, what had previously appeared to many as purely economic issues took on an overtly political dimension. When the dam of restructuring broke at the end of the 1970s, and now-surplus capacity began to be closed, the flood of protest which followed raised significant and difficult questions for both steel companies and, in many instances, national governments and states. In this turbulent period of conflict, many of the contradictions inherent to the growth path strategy of the earlier years, and to the character of state intervention, became clearly exposed.

We therefore focus in this chapter upon these anti-closure campaigns, examining decline by identifying how it actually came about; how, in other words, international restructuring revealed its political aspect. We commence with an outline of the contraction of the UK's main steel producer, the British Steel Corporation; the fortunes of the UK's private sector; and the prospective de-nationalization of BSC. We move on to consider in similar detail two further cases, France and West Germany, from inside the European Community. Finally, we briefly discuss the main trends from other EC states and the USA, in order to highlight the fact that politically contested closures were by no means exceptional in this period but rather had become frequent and often dramatic challenges to steel companies and national governments.

The United Kingdom

The British Steel Corporation to 1979

Upon its formation in 1967 with the nationalization of the fourteen largest UK producers in terms of output, the British Steel Corporation (BSC) was one of the largest bulk steel companies in the western world. Its early history reflected a period of organization and consolidation, before it embarked on a strategy similar to that recommended by the private sector before nationalization (BISF 1966). A £3,000 million investment programme (at 1972 prices) was to concentrate on the five 'heritage sites' of Llanwern and Port Talbot in South Wales, Scunthorpe and South Teesside in England, and Ravenscraig in Scotland (Department of Trade and Industry 1973), in an attempt to mimic the Japanese model of large coastal complexes (see Figure 3.1). Under this 'Ten-Year Development Strategy' annual capacity was expected to increase from 27 m tonnes to 33–35 m tonnes by the late 1970s and to 36–38 m tonnes by the early 1980s. One-third of the investment was to be allocated to South Teesside creating a giant, modern coastal works with an annual capacity of some 12 m tonnes. Associated with this new capacity, older plant at inland locations would be closed. Some of these cut-backs were delayed by the 'Beswick Review' initiated by the incoming Labour government in 1974, but some closures were

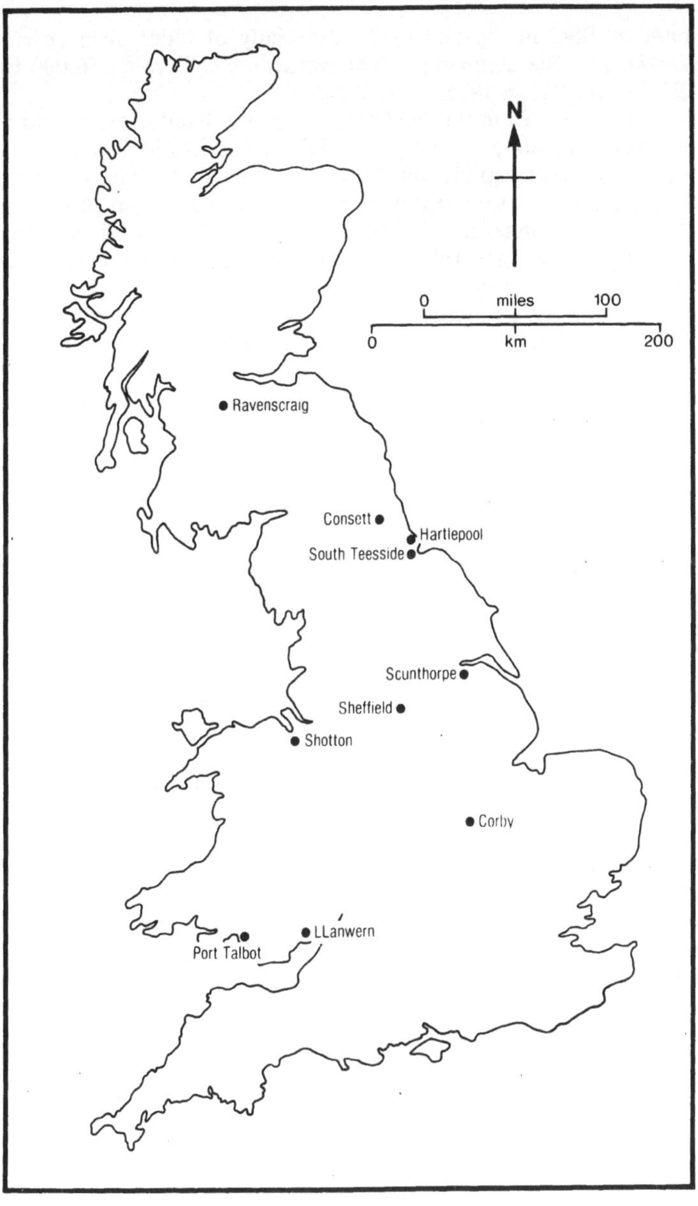

Figure 3.1 Selected steel towns in the UK

made at BSC in the mid-1970s, especially of older open hearth steelworks. The company's total workforce fell from 250,000 in 1971 to 210,000 in 1976.

Sadly absent from the 1973 scenario was a detailed examination of steel demand prospects, especially since crude forecasts of expansion were rapidly overtaken by events. As the recession deepened, BSC's steel production dropped from 25 m tonnes in 1971 to 17 m tonnes in 1976. Small profits were replaced by larger and larger losses (see Table 3.1). The drive to cut public expenditure which had been initiated by International Monetary Fund pressure in 1976 intensified, and BSC's policy of expansion switched to one of savage retrenchment. Formally announced in 1978 (Department of Trade and Industry 1978), the new strategy had in fact begun in March 1976 (Bryer *et al.* 1982: 171). At a time of rising unemployment nationally, closures became increasingly unpopular and were more fiercely contested locally.

During the early 1970s the trade unions in BSC had adopted a policy of active co-operation in closures. At its annual (non-policy-making) conference in 1973 the main steel union, the Iron and Steel Trades Confederation (ISTC), agreed to accept closures provided alternative employment was made available for redundant workers (Upham 1980). Such a response was widespread throughout the labour movement. For example the Welsh Labour Party adopted a similar resolution at its annual conference in 1975: 'no redundancies arising from the modernization of the steel industry should take place without suitable alternative employment being available within reasonable travelling distance of the workers' homes' (Nationalised Industries Committee 1977: III, 83). It seemed that if the price for cherished new investment was to be job losses, then at least the pain of transition could be minimized and, in any case, forecast output expansion meant that jobs at the coastal works would be secure for years to come.

The TUC Steel Committee (bringing together eighteen unions within a joint negotiating framework on any issue except wages) also reflected the accommodative nature of trade union organization in the steel industry, owing much to early traditions.[1] Its members justified inactivity over closures by a concern not to be seen to favour one works over another. This was expressed clearly in 1977 by Bill Sirs, then chairman of both the ISTC and the Steel Committee, writing about BSC's 1973 strategy:

> Painful change would have to be: this the Committee had to accept. But the Committee took a deliberate decision for which they were subsequently criticised in certain quarters, not to

Table 3.1 British Steel Corporation, 1967–88

	Profit (loss) (£m)	*Liquid steel output (m tonnes)*	*Capital expenditure (£m net of grants)*	*No of UK employees year-end (× 1,000)*
1967–68	(22)	22.9	—	254.0
1968–69	(23)	24.2	—	254.0
1969–70[a]	12	12.3	—	255.0
1970–71	(10)	24.7	—	252.4
1971–72	(68)	20.4	—	229.7
1972–73	3	24.2	154	226.6
1973–74	34	23.0	155	220.4
1974–75	70	20.8	273	228.3
1975–76	(268)	17.2	462	210.2
1976–77	(117)	19.7	494	207.9
1977–78	(513)	17.4	401	196.9
1978–79	(357)	17.3	267	186.0
1979–80[b]	(1,784)	14.1	261	166.4
1980–81	(1,020)	11.9	148	120.9
1981–82	(504)	14.1	164	103.7
1982–83	(869)	11.7	122	81.1
1983–84	(256)	13.4	164	71.1
1984–85	(383)	13.0	210	64.5
1985–86	38	14.0	220	54.2
1986–87	178	11.7	269	52.0
1987–88	410	14.7	253	51.6

Source: BSC *Annual Reports* 1967–88

Notes: [a] Six-month trading period from October 1969 to March 1970
[b] Figures affected by strike action, January–March 1980

> formulate a development strategy of their own. For they knew that if they did this those of their members at the works which any 'Steel Committee Strategy' would have to propose for closure might have felt abandoned by their own unions.
>
> (Sirs 1977)

Such attitudes fitted in with a policy of 'alternative employment' for redundant steelworkers, seen as the appropriate trade union response to technological change. This view persisted at least until 1977. Then, following the negotiation of enhanced redundancy pay by local union officials at Clyde Iron Works (in return for agreement on redundancies) the Steel Committee was called in to negotiate a slightly better deal at Hartlepool. From this time onwards the Steel Committee placed more emphasis on the negotiation of redundancy payments than on the provision of alternative employment (see Upham 1980: 8). The shift in emphasis did not disguise, however, a continued active acceptance of job losses and works closures on the part of national union officials.

On the other hand the Steel Committee was essentially useful to BSC in that it provided it with a framework within which the questions of job loss and works closure could be put forward (see Bryer *et al.* 1982: 243, 252–3). This was recognized by the Corporation when in 1975 it invited the Steel Committee to participate in the newly formed job-creation subsidiary, BSC (Industry) (see also Chapter 4). In accepting this invitation the Committee also implicitly accepted a degree of responsibility for closures. By the late 1970s therefore the Steel Committee had fulfilled a useful role for BSC management but was increasingly unpopular amongst steelworkers at plants threatened with closure.

These interrelated themes — the role of the TUC Steel Committee and national trade unions, and growing over-capacity as new projects came on-stream while UK demand slumped — came to the fore in 1979 when BSC proposed to close basic oxygen steelmaking at Corby, with the loss of 6,000 jobs. They rose to prominence at a significant political juncture, for the announcement of impending closure came under a Labour government in February of that year, but the campaign against closure was fought under a Conservative government newly elected in May and intent on reducing the state sector in any way it could. The way in which this campaign for the Retention of Steelmaking at Corby (ROSAC) developed sharply identified the paradoxes of previous policies.

Corby

The campaign in defence of Corby steelmaking was highly significant. It was the first basic oxygen steelworks to be scheduled for closure as a consequence of BSC's concentration on five coastal sites. Corby had grown around iron and steel since Stewarts and Lloyds started work in 1933 on its new plant to exploit local iron ores, and the jobs under threat represented a major blow to this one-industry town (see Ward and Rowthorne 1979). The campaign drew upon a broad range of local support for lobbies and marches to London and in the town (for a step-by-step account see Maunders 1987). At the same time arguments presented in favour of and against closure, and the policies of key parties, demonstrated the legacy of the now discredited Ten Year Development Strategy for the steel industry.

The main plank of BSC's case was that the tube works at Corby would be more efficient if operated on coil supplied from the South Teesside works. Here the giant Redcar blast furnace, part of the original expansion envisaged in 1973, had just come on-stream at a time when the Corporation already had surplus iron and steel-making capacity nationally. A financial analysis of BSC's proposal was undertaken during September and October 1979 by campaign advisers from Warwick University (for a summary see Bryer *et al.* 1982: 202–37). They submitted sixty-one questions to BSC management in this period and although BSC had previously agreed to answer any questions that might be put, in the advisers' opinion BSC completely failed to answer twenty-four of them. Despite this difficulty in obtaining information, the essence of the closely argued Warwick case was that savings from closing Corby and increasing production at Redcar were over-estimated by BSC, and that to close steelmaking at Corby and run Redcar required more capital investment than to make Corby viable as an integrated complex. This was placed in the context of a thorough-going examination of BSC's overall strategy, and its need for a financial reconstruction. In their response though BSC management disdainfully dismissed some arguments and completely ignored others.

The case against running the Redcar blast furnace did not just pose questions for BSC. It also struck at the heart of the trade unions' position on the need for new investment regardless of the consequences in terms of jobs. It left national union leaders in a dilemma, wanting to be seen to oppose closure at Corby but not by emphasizing the under-utilization of Redcar. This came to a head when the Steel Committee met BSC management at Corby on 20 September 1979. As the leader of the engineering workers' union (AUEW), Gavin Laird, argued that 'there's no point in setting off on the premiss that we don't commission other plants', BSC chief executive Sir Robert Scholey skilfully warned that 'without the elimination of fixed costs at Corby, cuts would have to be made at Redcar' (quoted in Maunders 1987: 181–3). Under pressure from all sides, even local union leaders asked the Warwick advisers to tone down the references to Redcar in defence of Corby (see Baker 1982: 52; Maunders 1987: 203).

Partly because of this (and partly in the knowledge of previous disappointments) a degree of suspicion characterized relations between local union leaders and the Steel Committee. At one point relations deteriorated to the extent that a representative of the Steel Committee was told that they were 'in cahoots' with BSC (quoted in Maunders 1987: 209). At this meeting, on 30 October 1979, local leaders also queried why references to the Consett works (and its impending closure if BSC followed the same logic as at Corby) had been deleted from the Warwick case. The answer, from Mr Delay, secretary to the Steel Committee, was revealing: 'Such an argument said "Goodbye Consett" and it was taken out because the Committee felt it was unfair to the workers of Consett who have not been consulted in any way' (quoted in Maunders 1987: 210). Not that this consultation was ever seriously on the agenda.

While the Steel Committee was feeling its way through the first months of a new Conservative government, Corby District Council was also doing its best to avoid entanglement with the campaign. At the May 1979 local elections control had swung decisively from Conservative to Labour. The newly elected council's strategy was one of distancing itself from the campaign in the belief that the fight would inevitably be lost, and that political controversy over the closure would be counter-productive in terms of attracting new industry to the town (see Maunders 1987: 101, 266; also see Chapter 4).

At national level too the Labour party was constrained in what it could do, not only by the magnitude of its parliamentary defeat but also by its earlier support for an expansionist programme (and, by implication, the running of Redcar). This was most clearly expressed by the Labour MP for Redcar, James Tinn. He made it clear that if Corby's possible survival would be to the disadvantage of his north-east constituents, then he would not be supporting it (Maunders 1987: 96).[2] The problem was deeper than just parish parochialism though, for nationally the Labour party was hamstrung by the fact that it was now (half-heartedly) opposing a strategy which it had already begun to implement in its period of office. Sir Keith Joseph, Secretary of State for Industry, made just this point in a House of Commons debate on the steel industry: 'The Labour Government, after shirking, flinching and shrinking from the social problems, accepted, towards the end of their period in office, that closures were inescapable if the new investment was to fulfil its purpose' (*Hansard* 7 November 1979: vol. 973, cols 446–7).

Faced with this lack of support the Corby campaign was ultimately a lonely one — although its final days coincided with a deepening recognition of the significance of Conservative government policies for the steel industry. In July 1979 Sir Keith Joseph had set BSC the target of breaking even by 1980–81, and significantly cut its external financing limit, the amount which BSC could borrow from government. Once again this mirrored the Labour government's policies for steel towards the end of its period in office, providing acute embarrassment for the party in opposition.[3] Financially constrained, BSC was moving to cut its losses by closures. Corby was in a sense the test-case and the manner with which BSC intended to push ahead with closures became very clear.

On 1 November 1979 the Steel Committee presented BSC with its detailed case against closure. The arguments were not even recognized (Upham 1980: 10–11).

This refusal to enter into discussion provoked a reaction from the ISTC. At an Executive Council meeting the following day the minutes record that 'despite the well-reasoned documentation that had been prepared . . . the Corporation were [*sic*] not prepared to countenance any alteration of their [*sic*] programme of closures regardless of the merit of our arguments'. A long debate ensued in which most contributors pressed for 'a more militant policy to be adopted'. In future negotiations 'the union should be prepared to exercise a greater degree of hostility' (ISTC *Annual Report* 1979: 227). However, the ISTC was unable to find support from other steel unions. At the November meeting of the Steel Committee an ISTC proposal for industrial action over plant closures was rejected (Upham 1980: 12).

In the face of this decision, the steel unions were soon drawn into a long and bitter dispute over wages. The first national steel strike since 1926 lasted from January to March 1980. In Corby it seemed that closures might also be an issue in the dispute. When it became clear that this was not the case, campaigners had nowhere else to go for support. The ROSAC campaign committee was formally disbanded on 3 January 1980. In the mean time though BSC had moved on and was proposing to close another steelworks as well — at Consett.

The 'Save Consett' campaign

The announcement that Consett was to close was made on 11 December 1979. Its future had been in some doubt for many years. In 1972 Peter Walker, Secretary of State for Trade and Industry, could guarantee only that Consett would operate as a steelmaking concern 'until later in this decade'. It was, he said, 'impossible to make a decision beyond that' (*Hansard* 2 December 1972: vol. 848, col. 1589). However, BSC's announcement still came as a tremendous shock to the town. A plate mill had been shut down two months previously with the loss of 300 jobs and it had been hoped that with this loss-making plant gone the works had a brighter future. Like Corby, Consett had developed as a one-industry town (see Sadler 1986: 136–9) and the 4,000 jobs under threat dominated local employment prospects.

A campaign developed to save the works, at first slightly confused by the start of the national steel strike on 2 January 1980. A small rally against closure on 8 February was followed by a larger one of some 3,000 people on 14 March. 'The Case for Closure' was presented by BSC on 12 June and was met by a further protest rally of some 2,500 on 20 June, a demonstration by some 600 Consett people in London on 9 July, and a trade-union commissioned document, 'No Case for Closure', presented to BSC on 23 July with a further protest march of some 2,000 two days later. BSC argued, though, that the trade union case presented no new evidence and on 4 September a meeting of 2,000 voted to accept closure and begin negotiations on severance terms. Eight days later the last cast of steel was poured at Consett (for details of the campaign see Hudson and Sadler 1983; Sadler 1986: 140–60).

Beginning in the context of a national steel strike over wages posed problems for local organizers at Consett. The craft unions there were initially opposed to strike action and this led to some later problems of co-ordination.[4] That the dispute was over wages (not closures) was a delaying factor in organizing a framework within which to oppose closure. Whether or not the strike was deliberately engineered over a wages claim in order to break trade union resistance to cuts in capacity (Routledge 1980), the final terms of settlement (a 15 per cent increase) were remarkably generous. After resolving this dispute, though, BSC was in no hurry to relieve the tension at Consett, nor to provide further information detailing the reasons for the proposed closure.[5] When they did, the campaign was entered into in earnest.

The BSC case initially rested upon a criterion of 'profitability' at works level. Consett's MP David Watkins had taken a deputation from the local authorities to meet BSC on 30 October 1979:

> we were told that Consett is not safe unless it is profitable and that BSC can guarantee absolutely nothing. We were told that we must talk about profitability and not about numbers employed.
> (*Hansard* 7 November 1979: vol. 973, col. 466)

In presenting such an argument, BSC could point to the poor operating record of the plant over the previous five years. The cumulative loss amounted to some £54 million. Unfortunately for BSC given this position, from September to December 1979 the Consett works made a small profit, a change due mainly to the closure of the Hownsgill plate mill. However, the BSC concern with profitability had served only to justify the closure in an attempt to mask the real issue, the need to cut capacity in response to a reduced external financing limit imposed on the Corporation. This was later recognized by David Watkins in an interview about the meeting with BSC Chairman, Charles Villiers:

> There is no denying we were quite deliberately misled. We were told at that meeting the Consett Works must be made profitable if it is to survive. That profitability was already in the process of being achieved, yet it was only a few weeks later that the closure of the whole works was announced.
>
> (*Consett Guardian* 13 March 1980)

And as Villiers himself later said of Consett: 'They did just get a profit but we had too much capacity elsewhere' (*Newcastle Journal* 19 March 1982).

The point was also made explicitly in BSC's document, 'The Case for Consett Closure', presented to local unions on 12 June: 'the Corporation proposed to close the Consett Works, since despite its much improved cost performance achieved in 1979, closure of Consett capacity will result in an appropriate reduction on the Corporation's overall billet capacity'. However the trade union reply followed a completely different line of argument: that Consett as a works was profitable and could remain so in the future, with an output of 240 tonnes of steel/man/year in 1979, above the BSC average. In so doing it overlooked the fact that BSC wanted to close Consett because of corporate over-capacity and did not even consider the reasons why BSC wanted to close capacity. In reply therefore BSC did not contest Consett's improved productivity but simply cited once again the need to cut capacity.

This concern with the profitability of Consett's steelworks took a further twist in the final episode, simultaneously tragedy and farce, of the campaign to save the works. On 2 September it was reported that a consortium headed by Mr John O'Keefe (managing director of Chard Hennessy, a Gateshead engineering firm) aided by Mr J. Carney (formerly of the 'Save Consett' Union Committee and who, it was announced, was to become Derwentside District's first Industrial Development Officer in October) wanted to take over Consett steelworks.[6] The consortium was reportedly named the Northern Industrial Group, borrowing the name, accidentally or deliberately, of an earlier grouping of capitalist interests in the north-east (see Hudson 1989: ch. 1). 'Exploratory talks' were held with Department of Industry officials in London (*Financial Times* 2 September 1980), although considerable mystery surrounded the identity of the consortium members: 'the search continued yesterday for the members of the elusive consortium' (*Newcastle Journal* 2 September 1980).

Seemingly no approach was made to BSC, however, who announced that unless a firm offer was made by 12 September, the furnaces at Consett would be allowed to die down. Four days after the last batch of steel was produced, the first details of the consortium and its plans became public (in the *Newcastle Journal* 16 September 1980). It allegedly consisted of ten (un-named) British businessmen who planned to buy Consett works for £3 million (BSC was asking for £100 million) and operate it with a workforce of 2,700, western Europe's highest productivity rate of 320 tonnes/man/year and a forecast first-year profit of £20 million.

Whatever credibility the consortium had did not last long. The names of its members were revealed after a fortnight of speculation in an announcement which severed their interest in the plant. Only two of the companies, Cronite Alloys and British Benzol Carbonising (with a combined turnover of £20 million), had shown any serious interest (*Sunday Times* 28 September 1980). O'Keefe's own company, Chard Hennessy, was revealed to have made pre-tax profits of less than £3,000 in 1977 and 1978 and its subsidiary, Potts and Sons, losses of over £50,000. More revealing still perhaps, given his prospective role as Chairman of the Northern Industrial Group consortium, was O'Keefe's revelation that 'I'm not really good at running companies. That sounds stupid, but I tend to get involved in the start of things and leave them to run themselves. If they don't do very well, I close them.'

Even the final bizarre act of the campaign, then, placed it into competition with other works as (supposedly) the most profitable steel plant in the country. At the same time, an anti-closure argument on profitability terms isolated the campaign from broader support either within the north-east or from workers at other plants within BSC. By reinforcing and replicating the claim for decisions to be made only within the context of one plant, the campaign failed to consider the broader issues in a fashion which would have invited support from other sources than just steelworkers at Consett.

At national level, after the steel dispute, the ISTC had finally prepared an alternative strategy for the steel industry (ISTC 1980). Its *New Deal for Steel* argued for a new aggressive commercial policy, a 10 per cent price cut, an improvement in product quality, and more customer choice of where an order was made within BSC. Too little, too late, it was scarcely relevant to the Consett campaign and fell on deaf ears in London — but it was symptomatic of where the national unions stood: prepared to argue that BSC should be more efficient without taking into account how 'cost' and 'efficiency' might be defined (see also Manwaring 1981).

With Consett closed, there was virtually no opposition to the closure of a third basic oxygen plant, Normanby Park in Scunthorpe, the following year. Manned production capacity was cut from 21.5 m tonnes in 1979 to 14.4 m tonnes by 1982, new chairman Ian MacGregor's 'Alamein Line'. BSC employment fell from 186,000 to 104,000. In the light of continuing over-capacity, however, even these cutbacks began to appear not to be enough to satisfy BSC. The hitherto unthinkable prospect of one or more of the five key sites being closed appeared on the agenda for the first time.

Ravenscraig: the political reprieve and its aftermath

Throughout August and September 1982 there was a spate of announcements of job losses at BSC, with continuing speculation over the future viability of the so-called 'big-five' integrated plants of Ravenscraig, Teesside, Scunthorpe, Port Talbot, and Llanwern. BSC's trading position had deteriorated sharply from the point where in March 1982 it was losing only £0.5 million a week to losses of £8 million a week in July and £7.2 million a week during August (*Sunday Times* 10 October 1982). In the face of this uncertainty the Strathclyde Steel Industry Group (bringing together regional and district councils, the Scottish TUC, MPs and MEPs) met in September to co-ordinate a campaign in defence of Ravenscraig. Under political pressure BSC chairman Ian MacGregor consulted Industry Secretary Patrick Jenkin, to be told that

> no decision on the closure of a major steelworks would be taken by BSC without close consultation and without the agreement of government. I am not going to shuffle off responsibilities for this on to Mr MacGregor's shoulders.
>
> (*Hansard* 22 October 1982)

During November the Scottish Affairs Select Committee of the House of Commons heard or received evidence from a broad range of bodies as part of an investigation into the steel industry in Scotland. Its report was published in December and concluded in no uncertain terms that a further rundown of the steel industry 'would be a disaster for Scotland — in industrial, economic and social terms'. The process of decline would be a 'grave threat to the industrial and social structure of Scotland' (Scottish Affairs Committee 1982: para 37). A Scottish lobby had effectively mobilized behind Ravenscraig.

A three-year reprieve for all the big five works was finally announced on 20 December by Patrick Jenkin:

> The government believe that it would be wrong to take irrevocable decisions on future steel capacity at a time of such major uncertainty. I am therefore asking BSC to prepare its plan for the next three years on the basis that steel making will continue at all five major integrated sites. . . . this does not imply that BSC will be required to maintain manned capacity at the current level of 14.4m tonnes, nor that all the facilities within each of the five major integrated sites will necessarily remain in operation.
> (*Hansard* 20 December 1982: col. 673)

The extent to which this decision not to close Ravenscraig was a *political* one, and the way in which the anti-closure campaign forced it openly to be seen as a political issue, was emphasized in a subsequent report from the Industry and Trade Committee:

> In the light of what we were told about prospects for demand and despite the Secretary of State's argument that there was 'great difficulty in being able to make any confident forecast . . .' we take the view that his decision to retain five sites was essentially a *political* rather than *economic* decision.
> (Industry and Trade Committee 1983: para 12; emphasis added)

While Ravenscraig was the prime candidate for closure in 1982 this was not officially confirmed by BSC until January 1983, when the prospective savings to BSC of closing Ravenscraig were clearly seen as the greatest of all options under consideration (see Industry and Trade Committee 1983: paras 219, 225). Ravenscraig was therefore not alone in its campaign. On South Teesside, the impending need to re-line the Redcar blast furnace (renewing its inner shell of refractory bricks protecting the steelwork from high internal temperatures) led to fears being expressed locally over the closure of this jewel in the crown of BSC's 1973 strategy (see Cleveland County Council 1983; Hudson and Sadler 1984; 1985b). Nationally the ISTC responded by calling a 24-hour strike in October 1982, and unsuccessfully picketed Immingham in an attempt to stop the import of steel. It was Ravenscraig that was universally and correctly considered the most at risk, though, and the broad base of support against such a proposal was significant in making it politically unacceptable (see also Sadler 1984). This was despite later attempts to justify the decision on the grounds of uncertain demand forecasts: they had never previously been regarded as problematic by BSC. The reprieve, however, was specifically only temporary and partial. In consequence continuing uncertainty was to surround the future shape of BSC.

This insecurity, especially at the three strip mills at Llanwern, Port Talbot, and Ravenscraig, reverberated throughout the following year. During 1983 lengthy negotiations took place between BSC and US Steel over a proposal to ship semi-finished steel slabs from Ravenscraig to the Fairless works in Pennsylvania. In the face of intransigent opposition from American steelworkers, and amidst fears that 1,200 jobs would be lost with the consequential closure of the Ravenscraig strip mill, the deal was abandoned in December. Announcing this decision new BSC chairman Robert Haslam stressed the continuous losses being made by the three strip mills, amounting to one-half of BSC's deficit of £2.5 million a week. The point was amplified early the following year when BSC chief executive Sir Robert Scholey told a House of Commons Select Committee that he could not see sufficient demand for more than two strip mills (Trade and Industry Committee 1984).

The debilitating and divisive effect of uncertainty upon trade unions in the steel industry had by now become very apparent. Llanwern union leaders launched a campaign to protect their works, to prevent it becoming a victim of Scottish political pressure to keep Ravenscraig open. They submitted evidence to the same Select Committee which compared Llanwern's profits to Ravenscraig's losses, and claimed far greater customer satisfaction with the quality of Llanwern output. The Committee finally recommended that, for the present, no works should be closed (Trade and Industry Committee 1984), but the damage to national trade union organization had been done.

In such unpropitious circumstances, steelworkers were called upon to support the National Union of Mineworkers, as the 1984–85 miners' dispute began in earnest. The legacy of bitterness and inter-plant rivalry, in the context of great pressure from BSC for further closures, led to a swift and almost total rupturing of relations between the coal and steel unions, as steelworkers fought (as they saw it) to save their works and their jobs. On 26 March production of steel at Scunthorpe was cut by one-half because of a coal shortage, down to around 30,000 tonnes per week. By the end of April it was reportedly down to 20,000 tonnes, the lowest level possible before a shutdown of steelmaking. During this period the South Teesside works produced record tonnages of crude steel, as its relatively secure access to imported coal enabled it to make up at least part of the deficit on Scunthorpe's lost production.[7] In late April the Ravenscraig plant was also affected when coal supplies were cut from two trainloads a day to just one. Within a week BSC responded by bringing in additional supplies of coal by road. At the same time, despite a 20 per cent cut in steel production, the

Llanwern plant ran seriously low on coal supplies and received an emergency delivery of four trainloads of coke. The deteriorating situation with regard to coal supplies led BSC Chairman Robert Haslam to warn 'If steel plants have to be closed because of the strike, perhaps some might never re-open' (quoted in *Financial Times* 10 May 1984). His message clearly put further pressure on steelworkers.

Even by late May production at Scunthorpe was still in the 20,000 to 25,000 tonnes per week range, with just two of the four blast furnaces in operation. During June BSC relied upon convoys of lorries to maintain at least some coal supplies to Llanwern, Ravenscraig, and Scunthorpe; and some iron ore supplies after the train drivers' union blocked delivery of ore shipments to Ravenscraig. Following the refusal of train drivers to cross an NUM picket line at the BSC terminal for Scunthorpe at Immingham, local dockers claimed BSC used contractors to load iron ore lorries (in breach of the National Dock Labour Scheme) and walked out. The following day, 10 July, a national dock strike began over the issue. This was resolved within a fortnight but ore supplies at Immingham suffered continued disruption. By the end of July, after a month of concerted effort by the miners to blockade the steel plants, Ravenscraig and Llanwern remained dependent on lorries for the supply of ore and coal, while Port Talbot and Teesside continued to enjoy full benefit of their deep-water harbours. The scale of the supply operation, along with continued increased output from Teesside, was such that BSC had restored production to pre-strike levels of around 200,000 tonnes per week.

The summer of 1984 represented a high-point in the threat to coal supplies to BSC works, and throughout it steel union leaders reaffirmed their determination to keep steelworks operating.[8] ISTC refusal to meet miners' requests for production cuts was clearly aimed not only at maintaining the safety of coke ovens and blast furnaces, but also at protecting BSC from loss of orders and the possible loss of jobs. Co-operation with local management in maintaining coal supplies was a crucial part of this. While weekly losses mounted during the coal strike, to £8 million during July (of which £5 million was attributed to the extra cost of the coal supply operation due to the dispute), BSC was unwilling to propose further closures for fear of alienating steelworkers' support. Consequently a long-awaited Corporate Plan became instead a brief statement of objectives, leaving the future shape of the Corporation a matter to be resolved after the miners' strike.

Discussions between BSC and the government after this dispute finished in March 1985 led to continuing doubts over the long-term future of Ravenscraig, especially as the end of the three-year reprieve drew nearer. Its supporters mobilized once again in its defence. In June the Scottish TUC held an all-party conference to launch a fresh campaign and George Younger, Scottish Secretary, made it clear that he would once again fight any attempts by BSC to close Ravenscraig. In August BSC's proposals were announced. They contained mixed news for Ravenscraig. Its hot strip mill was not yet to close: indeed, all five key sites were to operate for a further three years. Instead BSC could buy out and close down a fourth UK strip mill, the privately owned Alphasteel plant at Newport in south Wales, and acquire its European Community production quotas. On the other hand the cold strip mill at Gartcosh, twelve miles from Ravenscraig and a finishing plant for its output, was to be closed by March 1986 with the loss of 800 jobs.

This was widely interpreted as threatening the long-term future of Ravenscraig, since all its strip output would have to be shipped elsewhere for further processing, adding to costs. George Younger, crucially, announced his reaction: 'any decision on the future of Gartcosh is a matter entirely for BSC' (quoted in *Financial Times* 7 August 1985). He was not prepared to support the Gartcosh mill. Others were less easily satisfied and a cross-party lobby argued forcefully that to close Gartcosh was tantamount to condemning Ravenscraig. The Scottish Affairs Select Committee investigated the proposal and indicated its support for this view: 'with the large fixed costs of an integrated steel plant, Ravenscraig could scarcely be viable without Gartcosh at any level of demand; the closure of Gartcosh does therefore seem severely to prejudice the future viability of Ravenscraig within BSC' (Scottish Affairs Committee 1985: para. 19). In January 1986 though, government rejected the recommendation and confirmed the closure of Gartcosh. Workers voted later in the month to begin negotiations on severance terms.

Such was the position, then, before the general election of 1987. Some excess capacity resulting from the 1973 investment programme had been removed with the hotly contested closures of Consett and Corby, and the quiet demise of Normanby Park. Even this was not sufficient to restore capacity in line with demand, as the strip mills (especially at Ravenscraig) limped on from one threatened crisis to another. As well as these major closures there had been a programme of drastic cut-backs in manpower at the remaining works, with selective closures of rolling capacity. BSC's workforce was down to one-third of its 1979 total (see Table 3.1). Alongside these developments in the public sector aimed at benefiting wherever possible from BSC's decline; indeed, they were arguably deliberately designed to hasten BSC's retreat. By 1987 these had reached a point where only bulk steelmaking activities remained in BSC's hands. We turn below to how this had come about.

The private sector, special steels, and south Yorkshire

Special steels production has historically been associated with south Yorkshire, which has the largest concentration of private sector steel producers of any area in the UK. After 1981 the private sector played a new role in the UK steel industry (see Hudson and Sadler 1987b; Sheffield City Council 1984a). This was facilitated by the Iron and Steel Act, 1981, which had three main provisions. These allowed for the eventual liquidation of BSC; the sale of assets to or joint ventures with private capital by BSC; and an exemption from BSC's previous statutory obligation to provide the full range of steel products. On the one hand private contractors moved in to undertake many tasks at BSC works (see Fevre 1986). On the other, in an attempt to prevent private steel production being forced out of business by BSC's drive to increase profitability from a near-monopoly position, joint ventures were formed between public and private sectors. These were codenamed 'Phoenix' in an attempt to designate an industry rising from the ashes.

In the process a new structure emerged in the UK steel industry. Seven major completed Phoenix companies involved BSC in partnership with companies such as GKN, TI, Johnson and Firth Brown, and Caparo (see Table 3.2). These were created with active government support. Sheffield Forgemasters represented no saving to the public purse and Allied Steel and Wire cost more to establish than if the existing BSC businesses had been maintained (Public Accounts Committee 1985). Once formed, the companies were active also in introducing new patterns of industrial relations into the steel industry. Sheffield Forgemasters imposed new collective bargaining arrangements in 1985 which were bitterly contested during a sixteen-week dispute. United Merchant Bar was the first company in which BSC had a stake to introduce a single-union, no-strike agreement. But the most significant effect of reorganization into Phoenix companies was the closure of existing works. This was most apparent in the largest venture, Phoenix II.

Negotiations between BSC and GKN over Phoenix II, later called United Engineering Steels, commenced in 1980 but were not completed until 1986. Throughout this period other private-sector producers closed capacity, clearing the stage for the new joint

Table 3.2 BSC joint ventures with the private sector: the Phoenix schemes

Company	*Shareholdings*	*Products*
Allied Steel & Wire Ltd (formed 1981)	50% BSC 50% GKN (BSC also has 68% of preference shares)	Wire rod reinforcing bar and light sections and wire
Sheffield Forgemasters Ltd (formed 1982)	50% BSC 50% Johnson and Firth Brown (BSC also holds 50% of 13% unsecured loan stock)	Forgings
British Bright Bar Ltd (formed 1983)	BSC 40% GKN 40% Brynmill 20% (BSC also holds floating rate subordinated loan notes 1993 — 43%)	Bright bar
Seamless Tubes Ltd (formed 1984)	BSC 75% TI 25%	Seamless tube
Cold Drawn Tubes Ltd (formed 1984)	BSC 25% TI 75%	Cold drawn seamless tube
United Merchant Bar Ltd (formed 1984)	BSC 25% Caparo 75% (BSC also holds 100% of unsecured loan stock)	Light sections
United Engineering Steels Ltd (from April 1986)	BSC 50% GKN 50%	Engineering steels; closed die forgings

Source: NEDO 1986a

venture. Main markets for engineering steels, especially vehicle components and drop forgings, were in a state of severe decline. In 1981 Duport closed its works at Llanelli after heavy losses. In 1982 Round Oak Steel Works (formerly jointly owned by TI and BSC) was closed, along with the London Works re-rolling mills at Tipton in the West Midlands, sold by Duport to BSC in 1981. In early 1984 Hadfields of Sheffield was closed with the loss of 800 jobs as the company's parent, Lonrho, was bought out by BSC and GKN. F. H. Lloyd also shut its Dudley steelworks, leaving only BSC and GKN in the engineering steels sector.

In February 1984 BSC and GKN finally presented draft proposals to government. UK engineering steels capacity, at around 2.6 m tonnes a year, exceeded demand by 45 per cent. The plan embraced GKN's Brymbo works in north Wales, and four BSC plants in the south Yorkshire area. Approval was delayed, however, pending appraisal of the rationalization costs expected to be met by government.

Trade unions in the industry had known of the negotiations for some time and, fearing job losses, organized an attempt to prevent the merger. The ISTC warned that creation of a single venture would lead to an increase in import penetration as customers sought (through direct sourcing) to prevent over-dependence upon one UK supplier (see ISTC 1984). In early 1984 the ISTC executive approved a plan to impose production quotas on employers which would effectively 'freeze' production at existing levels so that work could not be transferred from any plant threatened with closure under the scheme. Sheffield City Council also expressed opposition to the proposed merger. BSC then announced, in March 1985, that it intended to close the 400,000 tonnes per year capacity Tinsley Park works, with the loss of about 800 jobs. An anti-closure campaign was organized, supported by Sheffield City Council, which produced a broad-ranging alternative plan for the works, but closure was confirmed in June by BSC and eventually agreed by the steel unions in August.

Closure of Tinsley Park, and the subsequent formation of UES, served only to highlight the extent of decline in south Yorkshire's steel industry. Employment had fallen from 60,000 in 1971 to 43,000 in 1979 and 16,000 in 1987. At the same time Phoenix II virtually completed a stage in the reorganization of the ownership structure of the steel industry. BSC was now confined to bulk steelmaking and a range of more profitable operations was in private hands. As BSC began to produce profits speculation increased over its future ownership. This was a question not to be resolved until after the 1987 general election.

Privatizing BSC

Detailed proposals for the privatization of BSC did not feature in the Conservative party's 1987 election manifesto. For a number of reasons, though, the sale of BSC very quickly came to appear on the House of Commons agenda. Balance-sheet profitability was on the increase in response to improving market conditions, after years of cut-backs and losses. Other elements in the re-elected government's privatizing schedule had developed complications; steel now seemed a suitable candidate to keep the momentum rolling. The groundwork was laid for the return of BSC to the private sector in October 1987, when Allied Steel and Wire, one of the first of the Phoenix ventures, was sold to a consortium of financial institutions and company management. The move was hailed at the time by Kenneth Clarke, Industry and Trade Minister, as 'an important milestone for the UK steel industry', since it reflected institutional confidence in the steel industry (quoted in *Financial Times* 23 October 1987). The impending sale of BSC was formally announced in December. In the debate which followed, a number of significant issues which had rumbled quietly for many years came once more to the fore; in particular, the future of the five main production sites.

Privatization was warmly welcomed by BSC management. Company chairman Sir Robert Scholey commented that it would give the 'freedom to deal with problems' including options which were 'not always politically acceptable' as a nationalized concern (quoted in *Financial Times* 4 December 1987). Such remarks were clearly directed at the future configuration of the company's steelmaking plant, and in particular the Ravenscraig works. A company announcement indicated that 'subject to market conditions' there would be a commercial need to continue steelmaking at all five sites for at least seven years. The government warned, however, that the Ravenscraig hot strip mill was surplus to requirements, and therefore guaranteed only until the end of 1989 (quoted in *Financial Times* 4 December 1987).

Such indications were bound to provoke reaction from Ravenscraig; in January 1988 this crystallized in a novel form. Motherwell District Council (in whose area the Ravenscraig works was located) and neighbouring local authorities commissioned a special report to investigate the best option for the form of privatization, including disposal of the company not wholesale (as had been foreshadowed by BSC and government) but in parts. One of these, it was suggested, could incorporate Ravenscraig with the coated strip and sheet mill at Shotton in north Wales (to which much of its output was delivered). Local MP Dr Jeremy Bray argued that this would create 'a very viable enterprise' (quoted in *Financial Times* 7 January 1988). This clear attempt to forge a new future for the Ravenscraig works by allying it in a new company structure with one of BSC's more profitable end-product activities prompted a fierce response: not only from the south Wales sites (which also supplied Shotton) but also from workers at Shotton, who had no wish to be drawn into what they saw as Ravenscraig's problems. Chairman of the joint union committee at Shotton, Vernon Kindlin, sent an uncompromising message to Ravenscraig: 'Back off and leave Wales alone. Stand on your own feet — we won't be part of your fight' (quoted in *Western Mail* 9 January 1988). The implications were apparent for all to see.[9] Condemned on almost all sides, the alternative proposals made little impression.

The character of trade unionism in the steel industry had been evolving over several years. BSC had actively developed new forms of wage bargaining, with a substantial portion of increases dependent upon locally negotiated productivity bonuses. New styles of labour relations and changed work practices had been central to the Phoenix companies, and were gradually extending into BSC. Just one example illustrates this tendency. At BSC's Hartlepool site in north-east England, only a pipemill remained following closure in the 1970s of two steelworks with the loss of 8,000 jobs. This pipemill employed 200 people on a short-term contract-only basis. A unique Enabling Agreement had been signed in April 1983 as an alternative to total closure, negating all other collective agreements between BSC and the TUC Steel Committee. The new contract of employment stipulated that maximum flexibility was required of every worker; any system, days, or shifts should be worked; there was no entitlement to redundancy or severance payment; and no minimum holiday entitlement, with holidays paid instead as cash on the completion of an order. Subsequently a number of workers received several continuous periods of employment across different production orders, but BSC refused to revise the employment agreement, arguing that each new contract represented a break in service. The local significance of this redefinition of working practices and employment conditions was considerable. In a context of severe long-term unemployment BSC was able to redefine its relations with trade unions and reorganize the production processes to its own considerable advantage, while insecure temporary employment became the best on offer to some of the town's potential labour force. Nationally the proliferation of such agreements epitomized the changing face of labour relations. Company paternalism also extended to share ownership for core workers. In February 1988 BSC announced that it wanted as strong an employee involvement as possible in the shareholding scheme of the privatized company. In many ways the company's remaining 50,000 employees had become actively allied with its managerial goals.

Preparing the financial market for BSC was the next task. An extensive advertising campaign was initiated by BSC, under the slogan 'In shape for things to come'. Clearly a number of matters remained to be resolved before a financial market unused to steel would commit itself, especially the company's own shape and prospects, and the price of the share offering. Publication of the annual report and accounts for 1987–88 in July 1988 confirmed its potential attractiveness to investors. Record profits of £410 million and a buoyant outlook led chairman Sir Robert Scholey to continue the campaign for early privatization. Flotation, he said, would 'take place at the earliest practicable date, which my Board colleagues and I very much hope will prove to be before the end of 1988' (BSC *Annual Report* 1988). From being a financial millstone around successive governments' necks, BSC had become a source of windfall profit. Its future now lay in wholly uncharted waters.[10]

France

Decline in the steel industry has posed similar problems of crisis management in France. As in Britain, construction and enlargement of coastal works left traditional steel-producing districts (in Nord-Pas de Calais and Alsace-Lorraine) facing an acutely uncertain future. The French state became increasingly implicated both in the construction of new capacity and in attempts to resolve employment problems in the interior regions. This intervention took place, though, against a background of differing political traditions from those of the UK. Up to 1981 there was no systematic attempt to increase the scale of public ownership in France. Governments had instead attempted to guide the economy through a series of National Plans (see Green 1983). Within these, steel was regarded as a strategic sector, and steel companies received increasing financial support from the late 1960s onwards.

This was particularly apparent in the construction of a major coastal steelworks at Fos, near Marseilles. In 1970 the French government agreed to provide nearly one-third of the estimated investment requirements and formed a holding company, Solmer, along with the second largest French steel company, to build and operate the plant. As costs escalated, further financial support was sought and in 1973 the two largest French steel companies each took a 47.5 per cent share in a reconstituted Solmer. The West German steel group Thyssen also took a 5 per cent stake and the French government contributed an additional Fr. 850 million in low interest loans (see Bleitrach and Chenu 1982: 162–3; Gwynne and Giles 1980). During this period too Usinor expanded the capacity of its coastal plant at Dunkerque from 3.6 to 8.0 m tonnes, with the construction of a giant blast furnace (see Gachelin 1980).

These new plants came on-stream just as demand fell. In 1975 Fos operated at only half its potential output and a planned second phase was later postponed (Hudson and Lewis 1982: 179–83). As over-capacity became increasingly apparent, the pace of restructuring accelerated. In 1977 16,000 job losses were announced. Such

Table 3.3 Usinor, 1973–83

	Crude steel output (m tonnes)	*Workforce year-end (× 1,000)*	*Profit (loss) (Fr. m)*
1973	9.1	40.9	164
1974	10.9	42.4	160
1975	7.9	41.5	(1,224)
1976	8.8	41.7	(1,245)
1977	8.3	48.0	(3,000)
1978	9.8	43.0	(2,492)[a]
1979	10.8	40.2	(993)[b]
1980	10.8	34.2	(1,229)
1981	10.6	31.6	(3,918)
1982	8.8	30.6	(4,987)
1983	8.5	29.3	(5,333)

Source: Usinor annual reports

Notes: [a] period 1.1.78 to 30.4.79
[b] period 1.5.79 to 31.12.79

Table 3.4 Sacilor, 1977–83

	Crude steel output (m tonnes)	*Workforce year-end (× 1,000)*	*Profit (loss) (Fr. m)*
1977	6.4	44.7	(2,283)
1978	6.4	34.8	(1,014)
1979	6.5	29.9	(1,367)
1980	6.3	24.4	(1,940)
1981	5.6	22.5	(2,786)
1982	5.0		(3,690)
1983	4.7		(5,276)

Source: Sacilor annual reports

measures, however, proved insufficient to stem mounting losses. The two major companies, Usinor and Sacilor, had accumulated deficits of Fr. 8,800 million over the period 1975–77 (see Tables 3.3 and 3.4) and the accumulated medium- and long-term debt of the steel industry had grown to Fr. 38,000 million by the start of 1978. With losses continuing during 1978 it appeared that the industry was on the verge of bankruptcy.

The French state reacted to this situation in October 1978 when it set in motion the *de facto* nationalization of the industry, though denying it was doing so. By means of a major financial restructuring the state took a direct 15 per cent stake in the three main steel groups. In addition to this, its holdings via the banks and major financial institutions effectively amounted to control over some two-thirds of the share capital of the three companies. Essentially the restructuring centred on a scheme to convert the companies' debts to the government (Fr. 9,000 million) and major banks, both private and state owned, (Fr. 9,400 million) into 'participatory' loans — that is loans on which a nominal 1 per cent interest would be paid for five years, in practice converting them from debts into assets to be added to the companies' capital. Holding companies were set up to control the production activities of the three groups (though in fact Chiers-Chatillon was merged with Usinor) which to all intents and purposes were under government control.

In return for this rescue from the verge of bankruptcy, the steel companies had to agree to draw up and implement, rapidly, an enlarged round of employment reductions and plant closures. In anticipation of, and in an attempt to defuse, reaction to these cuts, the French government announced in September 1978 the creation of a special industrial adaptation fund (the FSAI) with a budget of Fr. 3,000 million. It was intended to help create alternative employment in areas to be hit by steel closures; from it companies could obtain grants of 25 per cent of their investment costs plus low interest loans for another 25 per cent. The government claimed that it would create 12,000 new jobs. Unimpressed by promises of possible new jobs but certain of job losses, the response of the steel unions in Lorraine was to call a 24-hour strike on 29 September. Their fears were confirmed in December 1978 when details of the new closure plans were announced, including 24,000 redundancies in Lorraine and the Nord.

For a number of reasons this particular threat to steel employment in Lorraine and the Nord was greeted with massive protest which ultimately had national implications. Lorraine had faced economic decline before, with the rundown of its iron ore mining industry during the 1960s and 1970s. But the sheer scale of proposed redundancies, and their exceptionally localized character with grave consequences for particular localities, were of quite a new order. The timing of the announcement of the 1978 steel plan also provided the ideal platform for the opposition parties to demonstrate their particular policy line after a split between Communists and Socialists a year earlier. The Communist party, and hence the CGT, favoured a policy which included an increase rather than a rundown in steel capacity (see Gauche-Cazalis 1979: 33). In contrast the pro-Socialist CFDT preferred to accept some closures in steel, instead proclaiming the need for alternative employment in the affected regions (see Durand and Kourchid

1982: 90). And the increasing extent of state involvement within the industry had left government directly open to criticism of steel company operations within particular regions. Successive governments had not just participated in the development of chronic overcapacity but by 1978 were in overall control of the industry and therefore visibly responsible for its operation. Under these circumstances the incentive for opposition political parties and trade unions to attempt to discredit government policy on its own terms was particularly strong.

The days immediately following the announcement saw considerable activity in the town of Longwy where 7,000 jobs were at risk (see Figure 3.2). An inter-union co-ordinating committee, the *Intersyndicale*, was set up to fight the closures and link the actions of individual unions (Noiriel 1980: 32; Durand 1981: 83). To some extent, redundancies had been expected and prepared for. A huge SOS sign had been constructed, and was placed on the giant slagheap dominating the town after the announcement was made. A clandestine radio station began broadcasting in the area, with its title — *SOS Emploi* — reflecting the concern of its operator, the local CFDT. On 19 December some 20,000 demonstrated in Longwy against the plan.

In this early stage there was a fairly broad base of support against the closure proposals. For example, as early as March 1978 the local bourgeoisie had formed an organization, *Avenir du Pays Haut*, with the specific aim of protecting the region's economy (Noiriel 1980: 34). Moreover, the leading local newspaper, *Le Républicain Lorrain*, an institution of considerable significance within Lorraine (see Ardagh 1982: 59), very soon became actively involved in the protests. It published a petition each day after 23 December, calling on the President of the Republic to guarantee the future of Lorraine. This was sent to Giscard d'Estaing on 17 January 1979 with 40,000 signatures.

Further, 12 January saw a demonstration of concern within many parts of the region over the proposals. A 24-hour strike called by the iron ore mining and steel unions was extensively supported. A major demonstration was held in Metz, not itself directly affected by the plan, but seat of the region's prefecture; 60,000 people attended, while many similar smaller protest meetings were held in other towns. Transport and communications were also disrupted; trains between Paris and Luxembourg were stopped as well as all traffic into and out of Hayange and Rombas. The following day the newspaper *Le Républicain Lorrain*, in an editorial, described the day of action as going beyond class barriers; it did not represent 'a simple manifestation of solidarity at the level of one

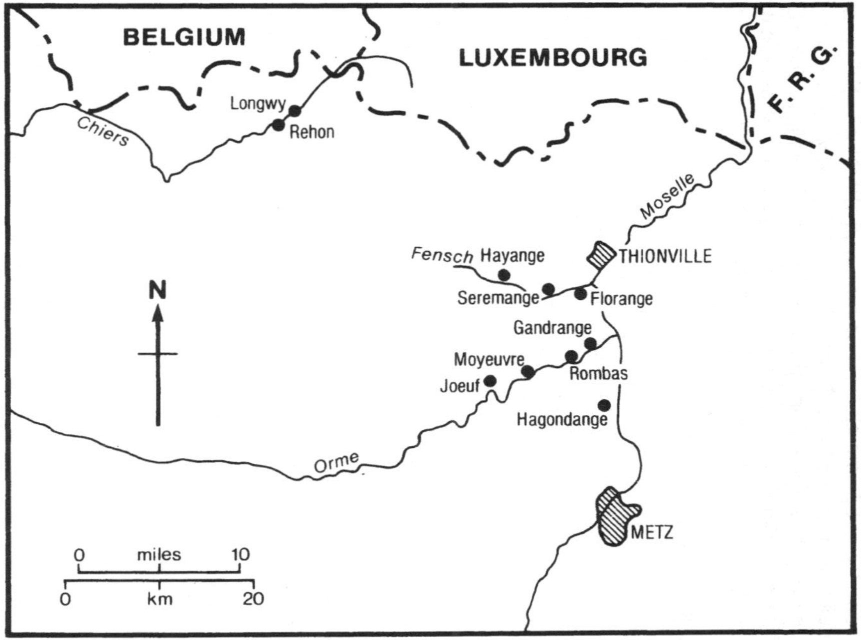

Figure 3.2 Main ironworks and steelworks in Lorraine

category of workers' but rather expressed 'the desire to struggle in order to continue to live and work in their region of birth or of adoption.'

Four days after this regional day of action, and in the fear of future protests, the government announced a plan to create 11,525 new jobs by 1982, mostly in Lorraine and the Nord. While the redundancies were certain, though, new jobs for steelworkers were not. In addition, for reasons that are not wholly evident, only 925 jobs were allocated to Lorraine (though there have subsequently been suggestions that the French government almost persuaded Ford to locate a car plant in the region, providing perhaps 8,000 jobs, and accordingly allocated little new employment there in the published plan; see Ardagh 1982: 60; *Le Monde* 24 January 1979). Perhaps predictably, the reaction in Lorraine was to regard the offer as a derisory one; indeed, the *Intersyndicale* at Longwy regarded the measures as a 'provocation' (*Le Monde* 20 January 1979). As a consequence, the steel unions called for a further strike within the region for 16 February.

Before these planned strikes could take place, however, protests erupted again in Longwy with direct action and civil disobedience growing in scale, such as the occupation of factories, banks, and offices connected with the steel companies. This extension of the protests into civil disobedience and physical violence, challenging the state's authority in these ways, saw the *Avenir du Pays Haut* begin increasingly to distance itself from the campaign. The emphasis on these means of protest reached new heights during the night of 29/30 January, when police forced entry to the Chiers works in Longwy to free local management who had been held there against their will by a group of protesters. The violence associated with the action drew a variety of responses: some condemning the police, others cautioning against violence, and in the process distancing themselves from the campaign.

Nevertheless, the second round of general strikes in Lorraine went ahead on 16 February, accompanied by major demonstrations; a marked escalation in direct action, blocking transport routes and disrupting trade and travel; and a national steel strike which received more or less total support. It appeared that the French government was losing control of the situation in the steel region, while the steelworkers were becoming increasingly confident of their own strength and capability to paralyse the region's economy. In an attempt to retrieve the situation, the government announced that it would stand firm on its plan for steel; not to do so would lead to an internationally uncompetitive industry and endanger the performance of the entire French economy. At the same time, protests intensified, most notably on the evening of 23 February and early 24 February when police moved in to halt the occupation of a local television station by protesters against the closures: those occupying it were objecting to the station's coverage of the campaign. This led to 2,000 more protesters besieging the local police *commissariat*, together with a bulldozer.

In March the most serious outbreak of civil disorder to date broke out, with rioting following the involvement of the French riot police, the CRS at a demonstration in the town of Denain, in the Nord; seven policemen were wounded by rifle-fire, thirty demonstrators by tear-gas. Later in the month, on the 23rd, a major demonstration in Paris of some 120,000 culminated in violent clashes between demonstrators and police in the Place de l'Opéra. Five days later the government and the steel firms agreed to review the closure programme.

The events of March 1979 were the most serious break-down in public order in Paris since May 1968 although seen retrospectively, they formed the peak of the protest. Moreover, it seemed for a time that the French state was caught in the grip of forces which meant it was unable to resolve the problems facing it. For it could neither abandon the steel plan nor increase the resources available for reconversion programmes in steel areas to appease opposition to this plan; hence the only way in which it could contain opposition was by physical violence and repression, which in turn only served to heighten resistance to the planned closures. Moreover, and perhaps even more seriously for the French government, opposition to the steel closures was becoming generalized into widespread protest against the whole deflationary tenor of its economic policies, and in particular against the continued rise in unemployment to which this led. This situation seemed to pose a threat at least to the authority of the French national government, if not the French state.

Yet only six months later, while protests against continuing steel cuts rumbled on, this major threat had disappeared. In part, this came about because the coherence of the protest movement itself began to break up. National strike action in steel began on 11 April, but at Fos and Dunkerque the issue was one of wages rather than job losses or capacity cut-backs; it ended in bitter defeat early in May. The result of this was further to heighten divisions between the various steel plants and steel unions over whether to fight the closures or campaign for alternative jobs.

Thus increasing divisions between the steel unions opened the door for the French state to defuse the threats posed by opposition to the steel closures, although at considerable cost. Essentially the French government stitched together a package of measures over a period of more than two years which eventually helped divide the steel unions and bought off mass protests against closures. There were two major elements in this. First, a long-term programme to create new employment. Second, and much more politically important in the short term, were the measures taken to cushion the effects of job losses on steelworkers. These fell into three main categories. The first of these was a special grant of Fr. 50,000 for any steelworker agreeing permanently to leave the industry, which was in addition to the usual redundancy payments which guaranteed workers sliding-scale payments starting at 75 per cent of their last pay during their first year out of work. In addition, for migrant workers Fr. 10,000 was added. In all, about 6,500 workers accepted voluntary redundancy on these terms (Ardagh 1982: 61), and it is not without significance that these grants were announced on 8 March 1979 in the middle of the Denain riots, by the Minister of Labour, M. Boulin. The second set of measures was concerned with compulsory early retirement: all those aged over 55 in the steel industry were retired on 70 per cent of their previous salary, while a large percentage of those aged 50–55 (particularly those in physically demanding jobs) were retired on 79 per cent of previous salary; in addition, a monthly minimum payment of Fr. 2,400 was set. Some 12,000 men fell in these two retirement categories (Ardagh 1982: 61). The third set of measures dealt with re-training: the 4,000 workers to whom they applied had the right to refuse two alternative job offers but on the third refusal their case was examined by a special committee and they could be made redundant. If they took a new job that paid less than their job in the steel industry, their former employer had to make up 60–80 per cent of the difference, while if the difference was 15 per cent or more, the person had the right to a grant of an extra Fr. 10,000. This clearly appeared a generous package to many French steelworkers for by March 1980 over 50 per cent of the workers that the closure programme wished to get rid of had left the steel industry and the programme still had fifteen months to run, until June 1981. In these terms, then, the tactics of the French government were successful.

In another sense, though, this very success was leading to other problems, for the massive costs of the restructuring programme were seriously exacerbating the very macro-economic problems which the closures were supposedly helping to solve. These costs arose in three main ways: Fr. 3,000 million for the FSAI; Fr. 10,000 million over the period 1980–85 for restructuring the financial position of the steel companies; and some Fr. 7,000 million on the various short-term, ameliorative social programmes. In all, this

Fr. 20,000 million was equivalent to about 50 per cent of the total French annual budget deficit for 1980.

Nor was the reorganization obviously successful. Continuing losses at Usinor and Sacilor mounted after 1978 (Tables 3.3 and 3.4). With the election of Mitterrand as President and of a Socialist government in May 1981, even the chairman of Usinor, M. Etchegarry, though a 'fervent advocate of private enterprise' had to admit that nationalization was the only way of meeting the companies' financial need (*Financial Times* 28 October 1981). Indeed, the widespread disenchantment with the previous Barre government was in no small part due to the effects of the earlier opposition to steel closures becoming translated into more general opposition to government policies. Under these circumstances, the steelworkers might have reasonably expected a better deal under the new administration: in fact they did not get one and the discrepancy between outcomes and expectations was important in reviving violent opposition to plans for further employment losses in steel. In February 1982 it was announced that while the government would turn its back on market forces as a way of rationalizing the steel industry, nevertheless Usinor and Sacilor would have to consider ways of achieving this goal. At the same time the government undertook a clean sweep of the top-level administration of the two companies, installing M. Raymond Levy as president of Usinor and M. Claude Dolle at Sacilor. In July 1982 the two companies announced plans for the period 1982–86, entailing some Fr. 20,000 million investment but, at the same time, a further 6,000 to 7,000 job losses, concentrated in Lorraine. These plans provoked further outbreaks of direct action: the headquarters of Usinor's special steels division was burned down.

Partly in an attempt to prevent such demonstrations of concern the government made available a further Fr. 500 million towards the end of July, through the creation by the two steel producers of 'reconversion companies' in the steel regions. Such offers, however, when seen from the point of view of the steel communities, missed the point and had not been the reason for helping elect a Socialist government: the situation as seen by them was neatly summarized by M. Jules Jean, the first Communist mayor of Longwy, when he wrote to Mitterrand: 'these plans are quite contrary to your thinking and your wishes' (quoted in *Financial Times* 14 July 1982). Even if this was an accurate analysis of Mitterrand's views (see also Ardagh 1982: 117–18), it was becoming clear that the new Socialist government was no more capable of halting the decline of steelmaking in Lorraine than its predecessor, nor of delivering the alternative jobs which it promised.

As a result, disillusionment with Mitterrand and his government deepened; as M. Galey-Berdier, Communist Mayor of Morfontaine, a small dormitory town near Longwy put it,

> I voted Mitterrand and I cannot hide the fact that I am totally disillusioned. Not only did the left promise to save what remains of the French steel industry. It promised to improve it. But it is applying the same policies as the right.
>
> (*Sunday Times* 30 October 1982)

While the steel communities had ultimately been willing to settle for increased resources to counter steel closures from a right-wing government, from a left-wing government the demand switched to the preservation of the industry but this over-estimated the room for manoeuvre open to the in-coming government and the capacity of a change of government to effect a dramatic change of policies within the context of a capitalist state. Far from nationalization solving the problems of the French steel industry, it constituted simply one moment in a continuing process of restructuring and rationalization which threatened work and life for particular steel-making localities. As realization of this deepened within the steel communities, protest against further cuts emerged with a demonstration in Longwy in September 1983 and a series of disruptions to production. In February 1984 trains were stopped by barricades near Longwy and the local Socialist party offices were attacked by steelworkers. On 29 March revisions to the original 1982–86 plan were announced, based on maintaining steel production at 18–19 m tonnes per year rather than increasing it to 24 m tonnes. A further 20,000 job losses were envisaged, mostly in Lorraine.

The new proposals met with instant response. The following day steelworkers clashed with police during a series of demonstrations in Lorraine with a wave of fresh civil disobedience, blocking road and rail traffic and ransacking the Socialist party offices in Longwy. The future of the Communist party within the ruling French government coalition appeared to be in doubt as the wave of militant action intensified, with the Communist general secretary, M. Marchais, calling the revised steel plan a 'tragic error' (quoted in *Financial Times* 3 April 1984). On 4 April Lorraine was brought to a virtual halt by a general strike called by local parties of both left and right and by all trade unions, while the Bishop of Metz and Nancy added the symbolic support of the church by allowing church bells to ring. Marches took place throughout most cities, involving 15,000 at Longwy. Care was taken to ensure protest passed peacefully but by the evening sporadic clashes had broken out between demonstrators and the CRS. Echoing the protest of 1979 35,000 steelworkers and supporters marched through Paris on 13 April, with the strongest support from the CGT. The atmosphere was summed up by a banner with the slogan 'Mitterrand, your electors are on the streets' (*Financial Times* 14 April 1984). The Communist party continued its opposition while speculation raged over its future in the coalition government. On 19 July it refused to join the government formed by the new Prime Minister, M. Fabius, because of the industrial policy. Thus finally the steel plan had the consequence of separating the coalition partners, the Communist party preferring condemnation of the cut-backs over continued participation in government.

The changing character of the Socialist government became even clearer the following year. In July 1985 a jointly owned subsidiary of Sacilor and Usinor, Unimetal, announced plans for an additional 2,000 job losses and the closure of its Trith-Saint-Leger plant near Valenciennes, which employed 770 people. This had been reprieved under former Prime Minister Mauroy the previous year; a prominent northern politician, he was also mayor of nearby Lille. After a bitter intra-party row, new Prime Minister Fabius pledged that the works would not close until alternative jobs were found for the 770 workers. Mauroy was appeased but four northern Socialist councillors resigned from the party in protest.

Even additional closures were not sufficient to stem mounting losses, and at the end of 1985 the French government was obliged to make a further Fr. 20,000 million available to the steel industry for the following two years. As disenchantment with the Socialist government grew, the 1986 election returned the right to power with a narrow majority. The following year a joint chairman was appointed to Sacilor and Usinor and the two groups were merged to facilitate further reorganization. A first step in this was the announcement of a further 16,000 to 17,000 job losses by 1988. In more than one sense, the political wheel had swung full circle.

Events in the French steel regions after 1978 thus represented expressions of opposition to closure with dramatic national implications. Drawing upon a reservoir of regionalist sentiment in Lorraine the steel plan of December 1978 was greeted with action across a variety of usual interest groupings, supported by many tens of thousands of workers and even drawing support from locally bound employers in the form of a temporary regional cross-class alliance. At the same time similar violent protests in the other major threatened steel area, the Nord, meant that the issue was raised in a cross-regional fashion, extending ultimately to massive demonstrations in the national capital symbolic of the wider significance of the steel plan. In the face of these circumstances the government was forced to delay the programme of closures and make available massive extra resources for steel communities, but the increasingly apparent inadequacy of these policy measures had the effect of translating opposition to the steel plans into general opposition to government policies.

A left-wing government coalition of Socialists and Communists was elected in 1981 on the basis of this dissatisfaction, which soon proved as unable as its predecessors to solve the steel regions' problems within the context of a capitalist state. Disenchantment within these areas grew upon announcement of further employment reduction in 1982 and reached crisis point in 1984 when local opposition to further downward revision of planned steel employment was translated by the Communist party into opposition to the overall tenor of economic policy, reaffirmation of opposition to the mode of policy formation and eventual resignation from the coalition government. The trade unions too signalled their increasing disillusionment which led ultimately to the return of the right to power in 1986, charged with yet further reductions in the industry.

West Germany

Even in a strongly free-market economy like West Germany, there have been considerable political problems surrounding the steel industry. The country is the biggest steel producer in western Europe and its output is dominated by five big private sector companies located in the Ruhr valley, West Germany's manufacturing heartland (see Figure 3.3; Sadler 1986: 248–55). Problems of overcapacity have been most apparent in the east of the Ruhr, away from its junction with the Rhine and access to the sea. In the eastern town of Dortmund, Hoesch was the biggest steel producer. Its recent history exemplifies the changing geography of production.

Hoesch became the dominant steel company in Dortmund in 1966 when it merged with DHHU. The 43 per cent share which the Dutch steel company Hooghovens held in DHHU was converted to a 15 per cent stake in Hoesch (Fleming and Krumme 1968). There soon developed a lively debate over future policy. The former president of DHHU, F. Harders, now vice-president of Hoesch, favoured the construction of a new coastal complex to compete on the world bulk steel market. This was not popular with all within the company administration, especially the president W. Ochel, who favoured instead the building of a new steel plant in Dortmund. In 1969 he was moved to the company supervisory board and

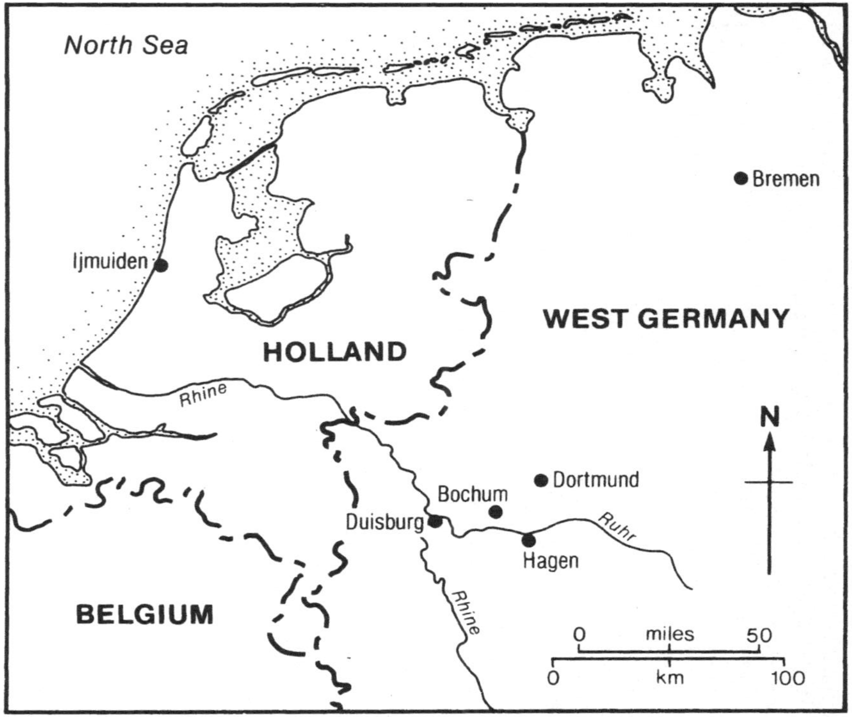

Figure 3.3 Steel towns in the Ruhr, Ijmuiden, and Bremen

Harders became president (see Schröter 1982: 13). However he was unable to raise the finance for a new coastal plant, and a total merger with Hooghovens came to appear as the best alternative, taking advantage of the Dutch company's coastal works at Ijmuiden. There was some opposition to this from the Dortmund workforce, fearing that the move threatened their employment. To appease this, the Hoesch management proposed to build a new steel plant in Dortmund, while Hooghovens mooted the possibility of a new works at Rotterdam.

Hoesch merged with Hooghovens on 1 January 1972 to form Estel, agreeing to take 300,000 tonnes of semi-finished steel annually from the Dutch concern. The following year Estel put forward a new strategy which entailed no new steel plant for Dortmund. Instead the existing open hearth works would be modernized for use until 1990. The Phoenix works, with oxygen steel converters, would also be closed after 1990. Despite speculation in 1974–75 over the possibility of a totally new plant on a new site in Dortmund, the strategy was confirmed in May 1979. Steel production at Dortmund was to be allowed to stagnate. Production costs of steel were DM 38 per tonne greater than at Ijmuiden: hence the Hoesch concern was to concentrate on steel processing as part of an attempt to stem Estel's mounting losses (see Table 3.5; Jäger 1981: 135; Schröter 1982: 17).

Table 3.5 Hoesch and Hooghovens, 1972–79

	Crude steel output (m tonnes)	*Steel division workforce (× 1,000)*	*Net profit (loss) (m guilders)*
1972	11.2	47.7	87
1973	11.6	48.6	170
1974	12.1	50.8	323
1975	9.6	49.8	(202)
1976	10.4	49.6	(69)
1977	9.4	47.6	(417)
1978	10.4	46.1	(288)
1979	11.5	46.0	(174)

Source: Estel annual reports

Not only was such a strategy unacceptable to the workforce in Dortmund, but also it was unpalatable to the management on the Hoesch side of the Estel concern (the two partners maintained separate production management organizations). Consequently in the autumn of 1979 speculation resurfaced over the idea of a new steelworks for Dortmund. On 23 January 1980 a concrete plan was announced in Düsseldorf for the construction of a new oxygen steel plant at Westfalen to replace Hoesch's three open hearth works, dependent on DM 240 million of state aid. The place and timing of this announcement are crucial, for it was not until 30 January, one week later, that the supervisory board of the whole Estel concern gave its approval to the proposal.

Although a concrete plan had now emerged, little specific action was taken, save for Hoesch applying for planning permission for the new works in September 1980. This lack of action encouraged further speculation by the autumn of 1980. Following a board meeting of Estel directors on 29 October, from which no decision was announced at the time, 1,500 Dortmund employees on two rolling mills stopped work on 31 October in a protest against the continuing uncertainty. The following day 800 shop-stewards demonstrated in front of the company head office, voicing similar complaints. Their worst fears were realized when on 4 November Hoesch chief Rohwedder informed the town council of Dortmund that the plan had, in fact, been shelved. In the opinion of the Estel directors, he said, 'the present situation and the great uncertainty over future developments do not allow, at the present moment, major strategic investments to take place' (*Ruhr Nachrichten* 8 November 1980). A scheme announced with the benefit of West German state aid had been cancelled by a joint West German and Dutch board of directors. This was unacceptable to both the Hoesch workforce and Hoesch management.

The month of November saw considerable discussion over the implications of cancellation, originating in particular from the *Burgerinitiative Stahlwerk Jetzt!* (the citizens' initiative for a steel-works now!). It represented a broadening of the protest from just steelworkers to include other interests. In the single most important event of this period, an estimated 70,000 people demonstrated on 28 November in Dortmund town centre in support of the slogan *Stahlwerk Jetzt!* The size of the demonstration showed clearly the extent to which concern had spread to include a variety of groups other than steelworkers.

This expression of popular concern was followed by a Dortmund *Stahlkonferenz* on 18 December, attended by some 150 politicians, representatives of Hoesch, and union leaders. This met again on 30 January 1981, both times with no practical result. However, the conferences did serve to defuse the popular protest of late November for the majority of the political representatives stressed the provision of alternative employment rather than the construction of a new steelworks (see also Bömer 1982: 44). It was not until February 1981 that protest action returned, now however largely confined to those directly engaged in steel production. On 5 February 1981 all 1,100 shop-stewards of the Hoesch works met to discuss future tactics. The most immediate consequence was a series of short strikes on 9 and 13 February. Hoesch management responded with a new proposal, entailing the construction of a new steelworks not as a replacement for the open hearth works but rather as a replacement of the Phoenix works (Bömer 1982: 53). Further details emerged on 7 May, when Rohwedder disclosed that the plant was intended not only to replace the Phoenix works, but also would bring about the entire closure of the Union and Phoenix sites: as a result the Hoesch Dortmund workforce would fall from 21,000 to 13,000. The full plan was tentatively agreed by the Estel board on 29 May.

The plan also marked a new phase in the development of Hoesch, for it was dependent upon a large degree of state aid. The extent of the required support only gradually became obvious. On 11 June 1981 Dortmund town council considered a proposal that Estel could afford to finance only DM 1,100 million of the total DM 2,700 million investment requirements. Largely in response to this request for aid, German politicians openly encouraged Hoesch to consider instead the possibility of co-operation with other German steel companies. This line of argument found favour with the Hoesch workforce, who felt that as a consequence of the internationalization of capital, Dortmund was in danger of complete closure. The Phoenix works council had voted on 8 May for a 'divorce' from Hooghovens. In June the *Burgerinitiative* also claimed that Hoesch should leave Estel as the Dutch company had invested in Ijmuiden at the expense of Dortmund.

Moreover, talks between Hoesch and Krupp had been going on since July 1980. In the context of Hoesch's inability to raise funds for a new steel plant within Estel, politicians aimed to limit the amount of financial aid given to steel, and to ensure such aid was given to a wholly German company. A DM 1,700 million four-year state aid programme for the whole steel industry was announced in September 1981. It soon became obvious that a Krupp/Hoesch merger would have no room for Hooghovens. In January 1982 the president of the Dutch company speculated that they might be looking for a divorce from Hoesch. This was confirmed when a corporate merger between the two German concerns Hoesch and Krupp was agreed on 4 February 1982. The new company, to be named 'Ruhrstahl', would not include Hooghovens.

It gradually became apparent, though, that the conditions created for Hoesch by a merger with Krupp were unlikely to be acceptable. The Krupp annual report for 1981, dated 25 March

1982, confirmed that the two companies had 'examined the possibility of opening up more potential for rationalisation through inter-company co-operation'. It went on to suggest that the two boards had agreed a

> restructuring concept in the steel sector, from which significant advantages will derive for us . . . through the better utilisation of pig iron and crude steel capacities in Rheinhausen for supplying steel to Dortmund and through savings in investment in the flat steel sector.

Krupp envisaged that Ruhrstahl would involve Krupp supplying steel to Hoesch for rolling at Dortmund. The degree of over-capacity in steel at Krupp meant that the development of a new works at Dortmund was unlikely, to say the least.

Hence although Hoesch had formally separated from Hooghovens in November 1982, a merger with Krupp was no longer likely. In this context the West German government commissioned three 'moderators' to prepare a report upon the future potential for rationalization in the steel industry. This drew an immediate response. The Dortmund quarterly assembly of IG Metall representatives, meeting on 13 December 1982, voted in favour of the nationalization of the entire steel industry. The significance of this local initiative cannot be over-emphasized, for IG Metall had consistently argued against nationalization. On 7 January 1983 a meeting of all 1,100 shop-stewards of the three Hoesch works voted in favour of the nationalization strategy, despite the fact that the very previous day Eugen Loderer, the ultimate head of IG Metall, had spoken out against it. Having failed to dissuade the Dortmund shop-stewards, national IG Metall policy adopted instead a compromise measure, arguing that government financial help should not be seen as a subsidy or interest-free loan, but rather as a form of direct capital subscription. In other words rather than complete nationalization, the giving of any government financial help to steel companies was to be regarded as taking a share in the capital of such companies.

This shifting of position was of significance in the light of the 'moderators report' of January 1983 (Bierich *et al.* 1983) which adhered to the conception that co-operation amongst companies was the way to solve the problem of over-capacity. The main feature of the proposed solution was the formation of two regional groups for the production of flat products and heavy profiles: a Ruhr group incorporating Hoesch, Klöckner, and Salzgitter on the one hand, and a Rhine group incorporating Thyssen and Krupp on the other. Considerable criticism followed, focusing particularly on the intra-group transport cost disadvantage of the Ruhr group, and the apparent advantages to Thyssen of co-operation with Krupp.

Just as the planned Hoesch/Krupp merger broke down, so too did the moderators' proposals. Hoesch almost immediately refused to countenance merger with Klöckner while the Thyssen/Krupp merger was aborted in November 1983 when Thyssen demanded a further DM 1,200 million from the state as a condition for taking over Krupp, on top of the DM 500 million already offered. Thyssen planned instead to cut its own capacity under a project 'Concept 900', by which capacity was to be reduced by 900,000 tonnes a month and the workforce by 8,000 by 1985. Another proposed merger, that between Krupp and Klöckner, announced in October 1984, also ran into trouble the following December when the government refused to provide assistance of DM 500 million. Any kind of solution in terms of co-operation between companies was not realistically practicable.

In this sense Hoesch was illustrative of a number of problems facing the West German steel industry — the need for state financial assistance before companies were prepared to merge, given the debts accrued by almost all steel producers and the continuing overcapacity in the industry. The problem was particularly acute in Dortmund where finance for a proposed new steel plant was not realized either in the framework of Estel, Ruhrstahl, or the moderators' proposed merger with Klöckner. The depth of the crisis facing steel producers, coupled with precedents set in the coalmining and Saarland steel industries, drew into question crucial features of West German society, notably the limited role of direct state assistance to capital. As part of this developing crisis, steelworkers in Dortmund called for nationalization of the industry, a dramatic step accommodated by a shifting of position by the national union structure.

Calls for nationalization are of even greater significance in Dortmund given the earlier internationalization of capital by Hoesch and Hooghovens. Conceived in a time of economic growth, the merger rapidly experienced conflicting pressures within national societies for the maintenance of employment. In the words of Hoesch chairman Dr Rohwedder:

> People used to ask why I had left Bonn to run a lousy company in Dortmund, but they were missing the point: the real pull was the idea of European co-operation. But then I saw the reality was something else, a huge financial burden, frustrations, national rivalries.
>
> (quoted in *Financial Times* 21 March 1984)

Sentiments shared by Mr Hooglandt, Hooghovens chairman:

> I think that the basis of the merger, as it was seen at the time, was sound. But the whole plan was drawn up in the context of growth. It's much easier to grow together than to shrink together . . . and the reality turned out very different to what we had hoped for.
>
> (quoted in *Financial Times* 21 June 1984)

Such comments serve to reinforce the severity of the problems posed for *national* states by an *international* crisis.

Other European Community producers and the USA

Cockerill-Sambre, the major Belgian steel company, was formed in 1981 by a merger of Cockerill Steel of Liège with Hainault Chambre of Charleroi. The new company was 80 per cent owned by the state, reflecting a substantial programme of aid for the steel industry during the 1970s. Its problems were amongst the most severe of any European steel producer and were intrinsically connected with the delicate balance between political parties, and between French-speaking Wallonia in the south (where its steelworks were located) and Flemish-speaking Flanders in the north. With a capacity of 8.5 m tonnes and a workforce of 25,000, Cockerill-Sambre lost BFr. 17,000 million in its first year.

After the downfall of the Christian Democrat/Socialist coalition government that year, a new centre-right coalition moved to impose economic austerity measures by decree. These included cutbacks at Cockerill-Sambre, which provided a focus of opposition for a number of one-day general strikes. In 1983 the Belgian government appointed M. Gandois to prepare a restructuring plan for the company. He was to play a substantial role in negotiating an accord with another steel producer, Arbed, in the neighbouring duchy of Luxembourg. The Belgian and Luxembourg governments met in November to discuss plans for co-operation between the companies. The negotiations also included Sidmar, the second largest Belgian steel producer, based at Ghent in Flanders. This was 51 per cent owned by Arbed with a minority share of 22 per cent owned by the Belgian government. Fearing further job losses, steelworkers at Liège in Belgium went on strike.

Agreement on a broad-ranging production and closure sharing deal was made in January 1984. The four steel regions — Liège, Charleroi, Ghent, and Luxembourg — were allotted specific roles. Cockerill Sambre agreed to close a section mill at Charleroi and a rod mill at Liège, while Arbed closed a strip mill in Luxembourg. For the medium term the two groups also defined a complementary investment programme.

The Luxembourg government had refused to accept any deal in which Arbed lost control of Sidmar, its most profitable plant, even though it needed financial support. In consequence the Belgian government put up BFr. 3,500 million to buy new shares in Sidmar, and a further BFr. 11,200 million in the form of long-term debt financing, but its shares were non-voting and its holding was limited to no more than 49 per cent, leaving operational control with Arbed. This provision of funds for steel-production in Flanders, imposed by the terms of an agreement with the Luxembourg government, did not go down well in Wallonia and led to angry demonstrations by steelworkers in Brussels. The rod mill at Liège (scheduled for closure) was occupied in protest.

The Belgian government then began in earnest on negotiations with trade unions over a plan for Cockerill-Sambre. In May the unions were forced to concede a plan for wage reductions and redundancies. Workers over 55 were to take early retirement while those who remained would effectively suffer a 10 per cent wage cut by missing out on increases to which they were entitled under Belgium's system of wage indexation. In return the government agreed a further financing package for the Cockerill-Sambre group. By 1986 its workforce was down to 15,000 and capacity to 4.5 m tonnes; but, re-elected with an increased majority, the government sought to cut a further 2,000 jobs in an attempt to trim losses.

Meanwhile in Luxembourg job losses at Arbed were highly significant nationally. Its workforce fell from 29,000 in 1976 to 13,000 ten years later, representing some 10 per cent of the country's jobs. This contraction was made with considerable financial help from a government anxious to minimize the effects of the steel crisis. As Arbed president M. Faber commented, 'we had to deal with an industry that had grown far too big for the size of the country. We were faced with how to avoid putting an unsupportable burden on the state' (quoted in *Financial Times* 19 November 1986). The cuts came through an even balance of early retirement, non-replacement, and voluntary redundancy. In addition the company created a special anti-crisis division which found jobs for up to 4,000 workers on a range of local and municipal projects. The state agreed to pay 20 per cent of its salary bill. In total the duchy contributed LFr. 8,500 million over the period 1983–86, more than LFr. 23,000 per head of population.

Arbed faced problems not only in Luxembourg and Belgium, but also in West Germany, where its subsidiary, Arbed Saarstahl, was also in crisis. This had led the West German government to provide DM 3,000 million in aid to the company in an effort to preserve some jobs and capacity: the Luxembourg government was understandably reluctant to support the company financially. Arbed Saarstahl's chief output was long products. In this market it faced intense competition from Italy.

The Italian industry was divided between the large, state-owned coastal plants run by Finsider, and the hundred or more private-sector, smaller producers concentrated around Brescia, north of Milan (see also Dunford 1988, 173–85; Moro 1984). The first major production cuts were imposed by Finsider in 1982 as the large Bagnoli steelworks in Naples was virtually shut down, temporarily, to avoid breaking European Community production quotas. The move led to widespread industrial unrest in the city. In June 1983 the EC demanded even greater cuts, amounting to 5.8 m tonnes by 1985, of which 4.8 m tonnes should fall in the public sector. This was highly unpopular and led to an angry public debate, made worse by the enforced temporary shutdown of the Finsider works at Taranto and at Cornigliano near Genoa for three weeks at the end of the year.

Complete closure of Cornigliano was averted the following year when 1,500 of the 5,000 jobs were saved as Finsider's subsidiary, Cogea, continued to make billets for sale to the private sector. The Bagnoli works too limped on, operating at only 50 per cent capacity because of Community quotas — in the politically charged atmosphere of Naples, the government made no attempt to close it completely. Both public and private sectors had skilfully negotiated within the EC quota system, the former in the fear of widespread unrest that would follow major closure announcements and the latter in the knowledge that creating regulations was not synonymous with enforcing them.

By 1988, however, the pressures on Finsider had become even more intense. Trading losses in the previous year amounted to L 1,680 billion, further increasing the deficit accumulated since 1980 to L 12,400 billion. Such a financial millstone forced the company to announce long-awaited restructuring proposals to an inevitably hostile audience of steelworkers, especially in Genoa and Naples. The plan's envisaged cut of 25,000 from the company's workforce of 75,000 met with fierce political and trade union resistance. Just as contentious was an intention to form two new operating divisions. One would group together the potentially profitable works and sectors, including the big integrated plant at Taranto. In this new company 9,000 jobs were to be shed by 1990.

The other would incorporate the greatest loss-makers. Not surprisingly perhaps, employees at the plants scheduled to become part of this second group, led by 3,000 workers at Bagnoli, expressed particular concern at the prospect of being sold off or eventually closed altogether, quite apart from the anticipated 16,000 job losses by 1990. A long-delayed restructuring programme had met with the anticipated response.

The political significance of steel closures has not only been apparent in the European Community. In many other countries steel has posed acute problems for governments of whatever political complexion. In the USA, for example, the first significant and intransigent opposition to steel closures in the 1970s was at Youngstown, Ohio. Several campaigns developed, first against the closure of the Campbell steelworks of Youngstown Sheet and Tube (from September 1977 until March 1980) and at two plants owned by US Steel and Jones and Laughlin (from December 1979 until July 1980). In 1977 a religious coalition was formed with the aim of promoting a buy-out of the steel plant through a workers' co-operative. The federal government subsequently refused to provide financial backing for the venture. The second round of closures led to a similarly unsuccessful buy-out project, though with greater attempt at workers' control, accompanied by a brief occupation (see Buss and Redburn 1983; Kourchid 1987).

Some of the lessons learned at Youngstown were later applied in Pittsburgh (see Deitch and Erickson 1987). The Tri-State Conference, an organization formed in 1979–80 in support of the Youngstown campaigners, reacted swiftly to US Steel's plan to close the largest blast furnace in the Pittsburgh region, 'Dorothy 6'. Over a period of several years a group of unemployed steelworkers and activists stopped demolition of the plant on three occasions, and even initiated the creation of a new public body, the Steel Valley Authority, with statutory powers of purchase over the steelworks. Unlike an earlier, successful workers' buy-out at Weirton Steel, under an Employee Stock Ownership Programme, US Steel was unwilling to become a creditor of the new authority, which was therefore unable to raise finance to buy the threatened works. Agreement over closure was subsequently reached in January 1986. At both Youngstown and Pittsburgh, then, in contrast to the European Community, opposition to closures focused on the possibility of generating alternative commercial futures for the threatened works rather than on seeking state support. By and large, though, anti-closure campaigns met with the same result: failure.

Summary

In this chapter we have reviewed and analysed a variety of attempts to contest steel closure proposals. The dominant characteristic that virtually all of them share, in a variety of localities and countries, is that they failed. In one town after another, steelmaking and/or steel-processing ended. Despite the variety of ways in which closures were opposed, the logic of capitalist production seemed to impose itself relentlessly on these places and people who lived in them. In these circumstances, ex-steelworkers continued to fight for the right to live, learn, and work in their localities but more and more this became a question of competing for alternative work in industries other than steel. Such 're-industrialization' policies and politics are examined in the next chapter.

Notes

1 Bowen (1976) identified three features of the early organization of labour in the iron and steel industry which played a significant role in the long-term development of the labour movement in the industry:

(a) The contract system of labour recruitment, essentially a social division of labour in which entrepreneurs contracted with middlemen, usually skilled workers, who in turn employed the unskilled labour required.
(b) The method of wage regulation by sliding scales, exemplified by that instituted by David Dale at Consett Iron Company in the late nineteenth century (see Carney and Hudson 1978). Under this arrangement, wages were cut by agreement at times of slack demand for steel production.
(c) The role of arbitration and conciliation machinery in the settlement of industrial disputes, symbolic of an ideology which propounded a harmony of interests between labour and capital in the industry. This conciliation machinery was exemplified by the Board of Conciliation for the Manufactured Iron Trade of the North of England, established in 1869 and still active in the 1920s.

2 Mr Tinn later expanded on this theme in the House of Commons:

> I believe that the investment in the coastal plant was right. . . . There is no soft option. The modern plants are on the ground. The heavy financial cost is part of the burden that the industry has to bear. These heavy costs cannot be met by under utilising the plant, understandable though the motives obviously are of those who see this as a possible easement of the situation elsewhere.
> (*Hansard* 7 November 1979: vol. 973, cols 490–1)

3 For instance in a House of Commons debate Sir Keith Joseph responded to remarks by the Labour MP Mr John Silkin in the following fashion:

Sir Keith Joseph: He maintains that the difference between the present Government and the Labour Government is that this Government are behaving like suburban bank managers . . . by imposing a date by which the BSC must reach profitability. Even that is wrong, because this Government is doing no more than fulfilling what we thought was a commitment of the last Government.

Mr John Silkin: No.

Sir Keith Joseph: No doubt the commitment might have been departed from but this is what my predecessor, the right hon. Member for Chesterfield (Mr Varley) said on 22 May 1978:

> The BSC must get its finances straight as quickly as possible. . . . Part of the Government's policy is that the financial objectives of the BSC should be to break even by the financial year 1979–80.
> (*Official Report* 22 May 1978: vol. 950, col. 1105)

We are now in the financial year 1979–80 and all that we are doing is asking the BSC to break even one year later — by the financial year 1980–81.

(*Hansard* 7 November 1979: vol. 793, col. 450)

4 On 27 December 1979 the 600 craft union members at Consett approved the statement that:

> We believe any strike at Consett in the present circumstances would be an act of utter folly which could only result in the BSC's proposal becoming an immediate certainty. The strike would enable BSC to escape from their present embarrassing difficulty in attempting to justify closing a profitable works. The Corporation will be able to claim that the workers themselves had brought about the closure by striking.
> (*Consett Guardian* 3 January 1980)

5 The problems such a delay presented to the campaign organizers were recognized by John Lee, secretary of the 'Save Consett Campaign':

> We have had no positive information from BSC since the closure was announced in December, and we think this has been a deliberate tactic on their part. We think they planned not to give us any information in order to demoralise and frustrate our members to the point where they would be prepared to accept any decision.
> (*Consett Guardian* 12 June 1980)

6 Mr Carney had apparently forsaken his earlier Marxist critique of capitalism (for example see Carney 1976) and was now actively promoting the virtues of capitalism as the only route to the salvation of Consett. It would seem that a dramatic conversion had taken place as what had formerly been seen as the problem became seen as the solution.

7 The Redcar plant had originally been designed to use coking coal from the Durham coalfield (see Beynon *et al.* 1986). The failure to expand iron and steelmaking capacity as intended in 1973, however, meant that there was sufficient spare capacity at the iron ore unloading terminal to enable coal imports also to be shipped through there.

8 Attempts by the miners' unions to block coal and coke supplies to steel works led to some of the most violent picket line disputes of the 1984–85 strike, notably at Orgreave. This revealed the extent to which the Thatcher government was prepared to go in using physical force to break the miners' strike via radically redefining the role of the police (see Beynon 1985).

9 The message was subsequently reinforced during the second reading of the British Steel Bill in the House of Commons by Barry Jones, MP for Alyn and Deeside, quoting a letter from the Shotton works joint union committee:

> We are convinced that any hiving off of the industry into smaller groups or units would not be in the best interests of the BSC as a whole or any works in particular . . . the future prosperity of the Shotton works [is] best served by our continued association within the BSC and not by fragmentation of the industry however well meaning these proposals may be.
>
> (*Hansard* 23 February 1988: vol. 1439, col. 217)

10 Since we completed this text, BSC was privatized in December 1988. By February 1989 major closures and job losses in the tin plate sector in South Wales had been announced.

Chapter four

Replacing steel jobs: state policies for re-industrialization

Introduction

The collapse of steel employment in many countries of the developed world and the failure of anti-closure campaigns have had a devastating impact in localities previously dependent upon the industry as a sole, or major, source of livelihood. This decline exemplifies the de-industrialization of many of the former manufacturing heartlands of the advanced economies (see for example Martin and Rowthorne 1986). It has been both a social and a political process, accompanied by the increased exploitation of labour. Large groups have become virtually excluded from the employed labour force: the old and the young, and those living in parts of the peripheral regions (see Bytheway 1987; Coffield *et al.* 1986; Hudson 1986). High levels of registered unemployment have acted to encourage acceptance of changes such as flexibility both within work, and in and out of work (see Purcell *et al.* 1986; NEDO 1986b). New patterns of industrial relations have been established both in the steel industry (see Chapter 2) and in other sectors (see Bassett 1986; Morgan and Sayer 1985). Traditional forms of employee rights and bargaining have been terminated or truncated, either by imposition or 'voluntarily' through self-employment or the creation of a precarious small business.

Through this transition, a number of questions to do with the relations between national and local politics have become clearly apparent. In response to local campaigns against closures and for alternative employment from those affected by decline in steel and other traditional industries, local and national governments have intervened with a range of policy measures seemingly in an attempt to replace old jobs with new. These policies are an integral part of the complex political process of change. They include retraining for those made redundant directly or indirectly as a result of closures, and encouragement of the creation of new jobs through support to new and existing businesses.

We therefore focus in this chapter upon some of the major re-industrialization schemes most strongly associated with steel closure areas in the UK and France. We commence by outlining the role of BSC (Industry) Ltd and its relationship to the growth of local enterprise agencies, drawing upon the example of Consett. We move on to consider the UK's Enterprise Zone scheme and its application in two other steel towns, Hartlepool and Corby. Alternative strategies to the Conservative party's philosophy have been followed in the UK in the 1980s, albeit tentatively; we consider one example here, from the 'steel city' of Sheffield. We then compare the UK experience in this field with that of the French steel region of Nord-Pas de Calais; and introduce the role of European Community policies with the same objectives and aspirations. Finally, we draw together these measures and evaluate their short- and long-term significance and effectiveness.

BSC (Industry) and local enterprise agencies: the case of Consett

British governments since the 1970s have financed job-creation subsidiaries attached to the major state corporations. The first of these was British Steel Corporation (Industry) Ltd (BSC(I)) set up under a Labour government in 1975.[1] BSC(I) by no means represents the sole attempt at reducing unemployment in steel-closure towns, but it is particularly significant in that it seemingly addresses directly the problem of decline as a response to local demands for work and employment. Such policies were frequently campaigned for by trade unions in the 1970s, with a questionable prospect of success even then. Their effect in the recessionary years of the 1980s is increasingly marginal, in terms of their stated objectives.

BSC(I) has a dramatic conception of its effectiveness in terms of creating new jobs, claiming to have helped to create some 70,000 since 1975. In that same period British Steel has shed over 150,000 jobs. These are certain losses, yet it is not clear how many of the 'new jobs' created are still in existence; how many are part-time or full-time; how many have been supported by both BSC(I) and other agencies, and are thus double-counted in claimed employment totals; how many pay a similar wage to that which they supposedly replace; and how many have been taken by ex-steelworkers. This imbalance between certain job losses and uncertain job gains is clearly reflected in continuing high rates of unemployment.

In 1984 BSC(I) handed over responsibility for job-creation to local enterprise agencies in its eighteen areas of operation, known thenceforth as 'Opportunity Areas' (see Figure 4.1). Such agencies have become increasingly significant in the UK and in this sense the

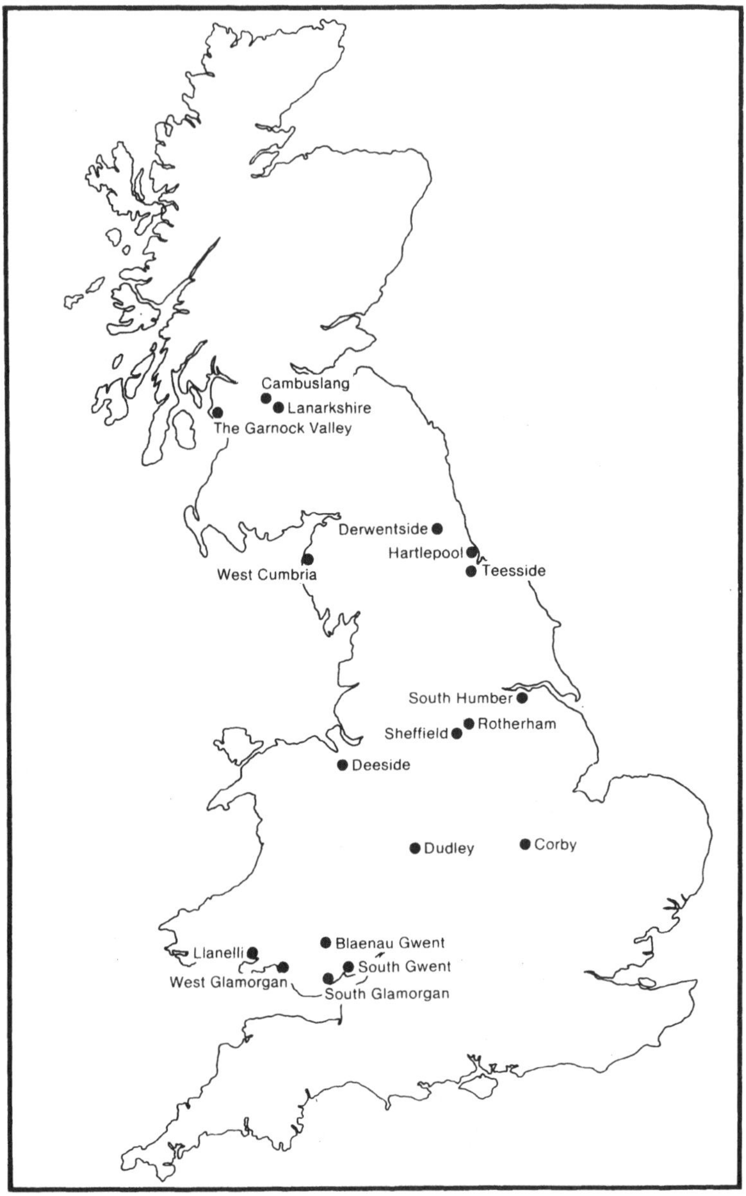

Figure 4.1 BSC (Industry) 'Opportunity Areas'

demonstration effect of BSC(I)'s example has not been without significance. Since 1979 local enterprise agencies have spearheaded the UK government's efforts to re-industrialize areas of particularly severe economic decline. The original model was the Community of St Helen's Trust, established in 1978 by Pilkington's (glass manufacturers) and the local authority. In the wake of inner-city unrest in 1981, their numbers mushroomed (with the support of Environment Secretary Michael Heseltine and Small Firms Minister David Trippier) to well over 200 by 1985. They epitomize the classic characteristics of the Conservative government's self-help philosophy. Local businesses club together with local government and major corporations in low-budget operations (typically £60,000–70,000 annually) which offer advice to potential new companies on business plans and financial incentives, and encourage the relocation of existing business to their areas. The responsibility for regeneration is effectively claimed locally under certain tightly specified preconditions.

In 1985, after a period of rapid growth, the first serious questions were asked about the long-term future of local enterprise agencies, especially with local authority budgets under severe strain and limits in sight as to what the private sector would contribute. In the light of fears that many faced serious financial difficulties, the Department of Trade and Industry investigated whether central government should extend its assistance. To that date it had contributed £2.2 million, mainly in the form of urban development grants from the Department of the Environment, although the Department of Trade and Industry's seven English regional centres were each able to provide £25,000 annually towards the establishment of new agencies.

As an expression of continuing political support, the government unveiled a five-year programme of financial aid in December, to replace the Department of Trade and Industry's assistance through regional centres. Government grants under the new scheme were to total £2.5 million in 1986–87, in addition to any separate funding made available through the Department of the Environment's urban programme. In each of the first two years of the new Local Enterprise Agency Grant scheme, qualifying bodies in England could apply for up to £20,000, to match resources generated from the private sector. Only agencies with an annual income of less than £60,000 could apply, to ensure the grants were targeted to smaller, more needy agencies. The total cost to the government was expected to fall over five years as each agency attracted a larger proportion of its resources from the private sector. To encourage this, the maximum grant was to decrease to £15,000 in the third year of the scheme, and to £10,000 in each of the last two years. For a marginal outlay, continuing self-promotion of the new business climate had been guaranteed.

The umbrella organization of the local enterprise agency movement, bringing together private sector sponsors, is Business in the Community (BIC), the main outcome of a working party set up in 1979 by Michael Heseltine, Environment Secretary, and chaired by Sir Alastair Pilkington. Its brief was to study what role the private sector could play in community affairs and local economic development. Judged by the dramatic expansion of enterprise agency numbers, these terms of reference have clearly been fulfilled. As David Grayson, a director of Business in the Community, put it: 'enterprise agencies became the macho symbol of the 1980s. Every town had to have one' (quoted in *Financial Times* 8 March 1988). Yet the objectives of members of BIC are many and varied. For some companies it represents a desire to create a social environment in which businesses can survive in the long term; for others it serves to deflect criticism when a large works closure provokes hostility. Since 1982 companies have been able to write off against tax contributions in cash or in kind to local enterprise agencies. For all it introduces a tension — whether or not their aim should be to maximize their balance sheet profit — and raises broader questions of the social responsibilities of employers. This ambiguity is often reinforced by the particular form of local enterprise agency organization and reflected in the character of local government involvement. This is clearly apparent in the experience of Consett since its steelworks closed in 1980.

Derwentside, encompassing the steel-closure town of Consett and the neighbouring town of Stanley, and still suffering from an earlier round of pit closures, has a high national profile. A succession of Cabinet ministers has come to pay homage to the job creation efforts in the town. As a traditional Labour-controlled local authority, Derwentside District Council has readily embraced the enterprise agency concept. Its association with local business represents a coalition of interests which is particularly reactive to comment, publicity, or criticism.

In 1984 BBC North-East produced a documentary programme, *Return to Consett*, which was broadcast in the region. It was based on a specially commissioned report which investigated the social and economic prospects of this part of County Durham (Robinson and Sadler 1984). In the period 1978–81, when the Consett steelworks and two major branch plants closed, employment in Derwentside District slumped from 28,000 to 19,000. Some 2,000 further redundancies have been announced since 1981. At that time the local enterprise agency, Derwentside Industrial Development Agency (DIDA), set up in 1982 to carry on the work started by BSC(I) in 1979, with its continuing support, claimed to have helped to create 1,800 new jobs. In 1980 BSC(I) had confidently predicted it would secure 6,000 new jobs for Derwentside within five years. This is what was said about the effectiveness of the local enterprise agency:

> Realistically, the *best* that can be hoped for from the re-industrialisation strategy is the creation of new jobs to compensate for those presently being lost — it cannot make up for the collapse of 1980–81. In other words whatever the merits of the re-industrialisation strategy, the starting point for thinking about the future for the area must be that high unemployment — in excess of 20% — will continue into the future.
>
> (Robinson and Sadler 1984: 80)

The DIDA reacted sharply to such comments. Its chief executive, Laurie Haveron, described the report as 'ill-informed pessimism, largely based on opinions rather than fact'. He went on:

> the authors have little realism to offer and the unending gloom depicted by them will hardly help us in our continuing drive to attract inward investment . . . there have been some notable jobs successes and we intend to continue building on them.
>
> (*Newcastle Journal* 25 May 1984)

As well as an apparent concern for image, there was a resentment at alternative commentary: the report, he argued quite simply, 'hasn't understood the real progress, in business terms, which is being made' (*Consett Guardian* 31 May 1984).

Several years later another paper compared the experiences of Consett with those of another English steel-closure town, Corby (Boulding *et al.* 1988; see also pp. 116–121). In this we called for an independent jobs audit into agencies like DIDA, which was reported in the local press (for example *Newcastle Journal* 22 May 1986). Our argument was a simple one: that with relatively poor official statistics (the census of employment is no longer annual, for example, and in any case is no longer released for the precise area of Derwentside which had enabled historical comparisons to be made before 1981) and with the only jobs information now available being that produced by DIDA (whose interests were in showing how successful it could be), no real impartial basis existed for a serious debate on the effectiveness of policy. This produced a heated reaction. DIDA chairman Mr Crangle complained that we were conducting our research in 'an ivory tower', and reassured the general public that the figure of jobs created was 'hard fact, involving no speculation or assessment'. He went on: 'It is deeply disstressing to see work like this' (*Newcastle Journal* 29 May 1986).

We felt that such comments should not pass unchallenged so we wrote publicly to the same newspaper. In our letter we expressed dissatisfaction with Mr Crangle's bland assertion that jobs created were 'on the ground', and asked for some further simple details about how many were full-time and part-time; how many were jobs transferred to Derwentside from elsewhere; what assessment was made of jobs lost in companies; and how DIDA defined the 'support' it had to give before a job created was its responsibility (*Newcastle Journal* 9 June 1986). These comments were clearly hard to answer at DIDA. In a reply Mr Terry Hodgson, then chief executive of Derwentside District Council (a sponsor of DIDA), could answer only one question — that concerning part-time jobs. He preferred instead to assert DIDA's superior competence in 'assessing the effectiveness of the agency's work', adding that 'it is a matter of regret that the once respected name of the University of Durham as a research institute has been associated with a piece of "work" which is so patently flawed' (*Newcastle Journal* 17 June 1986). Not content with such public outpourings, another of the agency's sponsors, the Derwentside Industrial Group (DIG), wrote to the Vice-Chancellor of Durham University (with copies to ourselves, Mr Crangle, and Mr J. Carney at DIDA). In his letter Mr Cockerton, chairman of DIG, informed us of his opinion that 'ill-founded, unconstructive research by your university, can only reflect badly on the good reputation of your university in general'. We, of course, entirely refute his emotive and unsupported assertions which were directed to deflect attention from the real point at issue: DIDA's effectiveness in creating alternative employment.

Clearly the members and sponsors of DIDA felt concerned about their public image and were more concerned to comment on this than whether the re-industrialization strategy was capable of making significant in-roads into mass long-term unemployment. One body which was not afraid to comment on this more central issue, though, was a joint working party of the Associations of County Councils, District Councils and Metropolitan Authorities (Local Authority Associations 1986). In a survey of local authority responses to decline of major traditional industries, the working party (which included in its number Mr Bill Hetherington, Chief

Planning Officer of Derwentside District Council) considered a number of case studies. One of these was Consett. This is what it said:

> Already there are signs of disillusionment arising in the population throughout the District. Although public morale has been sustained at a high level, social pressures do seem to be emerging. . . . We are running out of new ideas. Having worked very hard in the last five years we have in fact stood still in terms of unemployment because of other redundancies. The task of industrial regeneration is now looking to be very long-term indeed. . . . At the time of the closure it was thought that perhaps 10–15 years would be needed to re-create the industrial base of the District, on a much broader basis, in order that there would be an assured future. With the continuing international recession it seems that it will be certainly a much longer time period than this in order to achieve this regeneration. The big question will be whether the town and even the District can hold its population levels long enough.
>
> (Local Authority Associations 1986: 51–4)

As well as questions concerning the quantity of new jobs, it is significant also to consider their quality, in terms of employment conditions and contractual security. Many new jobs in Derwentside are low paid, highly exploitative and, in sectors such as textiles and clothing, often taken by women searching for any paid work because their husbands are unemployed. It is not uncommon for the activities of failed companies to be taken over by new companies, with questions often unasked and unanswered about debts and government grant payments. In Derwentside this problem has left workers facing an uncertain future on a number of occasions.

In 1985 premises at Tanfield Lea greeted the fourth company to start trading there in two years. The first, Fab Fashions, had gone into liquidation in December 1983. Another, Fair Fashions, was set up almost immediately but closed a year later, and a company called Quatriz Ltd opened (*Newcastle Journal* 17 October 1985). This closed in August 1985 and forty-five workers, mainly women, lost their jobs. As Derek Cattell of the National Union of Tailors and Garment Workers explained:

> Our people got no notice at all of the closure until the day it happened. They are owed wages in lieu of notice, outstanding pay and outstanding wages. No liquidator has been appointed

> and until that has been done none of our members can get any entitlement.
>
> (quoted in *Newcastle Journal* 21 August 1985)

As the workers successfully went to an industrial tribunal to win compensation, the fourth company on the site in two years began trading, with its managing director (mindful of past events) stressing he had 'no dealings with the previous tenants' of the factory (*Newcastle Journal* 13 November 1985).

Nor was this an isolated incident. Quatriz Ltd was only one of four Derwentside clothing firms taken to industrial tribunals by the National Union of Tailors and Garment Workers in 1985. Another of these was Tab Court Ltd, a clothing firm opened in December 1984 after a different business at the same factory, Tantobie Textiles, had gone into liquidation. Tantobie Textiles had subsequently been taken to a tribunal and the workforce awarded ninety days' pay in lieu of notice. Tab Court's managing director, Mr Whittaker, had been involved in Tantobie Textiles, and when Tab Court closed, the same problems ensued. The forty-six women workers were also forced to go to an industrial tribunal, and were awarded thirty days' pay in lieu of notice. The average gross wage of the workers had been £73 a week (*Newcastle Journal* 10 December 1985). Their predicament contrasted strongly with that of accountant Mr Burnard, a founder director of both Tab Court and Quatriz. At a public auction selling off machinery from the two companies, he commented that 'as far as I am concerned, we have got our fingers burned and we don't want to come back. It is a business and we are in it to make money. We wouldn't finance this type of thing again'. He added, however, that 'since last Monday, I have formed at least fifteen companies' (quoted in *Newcastle Journal* 1 November 1985).

The UK Enterprise Zone scheme: Hartlepool and Corby

The Enterprise Zone scheme was of considerable political significance in the UK in the early 1980s. The first zones were designated in 1981 with a second round following in 1983. Many encompassed areas of severe industrial dereliction, including several former steelworks and steel towns. The financial attraction of a ten-year rates holiday made them particularly lucrative sites for private investment. In their commitment to minimal regulation within the workplace, Enterprise Zones also embodied many of the salient features of contemporary UK government policy. The emphasis was on workers pricing themselves into a job, with little or no security, and on flexible working practices, backed up by the spectre of high national levels of unemployment (see Hall 1981; Shutt 1984).

Within a short time, Enterprise Zones came under considerable criticism on the basis of cost-effectiveness. Monitoring of first-round Enterprise Zones concluded that financial incentives were more significant than reduced bureaucracy, and that many zones had benefited from short-distance moves by existing firms at the expense of the surrounding area. While the zones provided significant potential for economic development, the measures in themselves were insufficient to guarantee this (Roger Tym and Partners 1984). By 1985 it was estimated that the 23 first and second round zones in Britain (two in Northern Ireland were monitored separately) had been established at a public cost of some £180 million. This comprised £50 million on rates foregone; £50 million on allowances on capital investment; and £80 million on reclamation and the provision of infrastructure. By October 1984 the first round zones had experienced an increase of 14,000 jobs, though many of these had to be discounted against moves from nearby. While it was early to make any judgement, the Audit Office concluded that 'the information available strongly suggested that the Exchequer cost per net new job created in the first round zones could be disproportionately high' (NAO 1986: 11). Partly in consequence, David Trippier, now Secretary of State for Inner Cities and Urban Development, announced in December 1987 that the government 'do not intend to introduce a general extension of Enterprise Zones' (quoted in *Financial Times* 23 December 1987). The policy focus had clearly shifted elsewhere.

Two examples of Enterprise Zones from steelworks closure areas in contrasting regions of the UK are Hartlepool and Corby. Within north-east England Consett is not alone in feeling the effect of steel closure. The town of Hartlepool saw similar cut-backs in the 1970s, earlier than Consett, and has an even longer experience of policy attempts to introduce alternative employment. It also remains one of the most obdurate unemployment 'blackspots' in the UK (see Morris 1987). In this environment many of the trends which are emerging in Consett are well-established in Hartlepool; features of the labour market such as low pay and insecurity are commonplace, and union activity is frequently openly discouraged. Unlike Consett, Hartlepool also boasts an Enterprise Zone, created in 1981 (for details see Boulding 1988).

Some of the larger new firms established on the Enterprise Zone have been hostile to any hint of trade union organization. PMA Textiles (the 'PMA' stood for 'positive mental attitude') interviewed 3,500 people for the first 24 jobs. Successful applicants were interviewed up to six times at various hours of the day and night. Employees were 'associates', prepared to work flexibly at any point in the 24-hour, 365-day process — though the normal weekly pattern was four 12-hour shifts. At another company, CDL 44 Foods, union activity is expressly forbidden and punishable by dismissal. Worker representation is limited to a consultative works council. Any employee arriving more than five minutes late on any one day is automatically sent home before starting work, and loses any attendance or productivity bonus for that week. In the first year of operation the owner sacked the entire workforce over the long Christmas and New Year period, re-employing them from 2 January, thereby obviating the need for paying any holiday money. The first factory to open on the Enterprise Zone in 1981, Lab Systems Furniture, is also non-union, as is one of the newest, the Singapore-controlled Tolaram. Management at Lab Systems Furniture maintains that profits are too marginal to offer wage rates comparable with the older employers in the town.

High unemployment, low wages, and new employment conditions are by no means confined to north-east England. Corby, which suffered closure of its steelworks in 1979, was also designated an Enterprise Zone in 1981. It provides a revealing comparison with elsewhere in the UK. In Corby businesses on the Enterprise Zone employed 3,600 in December 1985, down from 4,100 the previous year principally as a result of drastic reduction at one major branch plant (Department of the Environment 1985). In the town as a whole, according to a survey by Corby District Council, employment in manufacturing (except the tubeworks) was still on an upward trend, from 8,900 in September 1985 to 10,300 in September 1986. This gradual employment gain helped to reduce unemployment rates from a peak of 23 per cent in 1981, to closer to the national average.

The extent to which public support is essential to attract private investment is a significant question, especially for an Enterprise Zone situated in a relatively prosperous region of the UK. It can obviously be answered only with a high degree of uncertainty but the Audit Office tentatively suggested that in the first round Enterprise Zones the public:private investment ratio by 1985 was of the order of 1:1 or 1:2 (NAO 1986). This is hardly resounding testimony to the success of the pump-priming function of public investment. On the other hand the figure could well be weighted against such a conclusion since the zones, almost without exception, received heavy initial public investment to reclaim the land from previous (often heavy industrial) uses. To equate this with private investment in plant and machinery is misleading, especially since there is likely to be a longer lead time on private investment.

Some further indication can be obtained from a questionnaire survey administered in Corby during 1985 and 1986 (see Figure 4.2; also see Boulding 1988). One question asked whether the company would have established in Corby if the *complete* package of incentives (that is, not just the Enterprise Zone) was *not* available. Not surprisingly, perhaps, 86 per cent of the unequivocal responses from companies established prior to the steelworks closure in 1979 felt that they would still have established in Corby (and some of the 'don't know' responses included long-established companies to whom the question would be hardly relevant). By contrast, 64 per cent of the clear replies from businesses newly established after 1979 indicated they would not have established in Corby without the package of incentives. Among companies with more than ten employees this trend was even stronger, up to 76 per cent of those responding.

A second question asked whether the financial assistance available had been crucial for the survival of the business or plant. Once again the 'old' businesses followed the expected pattern, 88 per cent of them responding in the negative. This is of relevance in that grants on replacement investment are available to old and new companies alike. Revealingly a small majority (56 per cent) of the new businesses also answered that the assistance had not been essential for their survival. This was particularly apparent from the businesses employing ten or fewer people, of which 76 per cent responded in this fashion.

What emerges tentatively from the responses to these two questions, then, is that for new businesses (and especially for those with ten or fewer employees) the package of incentives was an important factor in their decision to locate at Corby (rather than anywhere else) but that these incentives had not in themselves guaranteed the survival of these businesses. If business survival is not guaranteed by the incentives, what are the longer-term prospects for these companies generating employment growth? A third question in the survey asked whether the effects of financial incentives had been to increase the number of people employed. Conclusions from this question are of more limited significance because of the larger number of 'don't know' type responses (in itself an interesting phenomenon). The clearest indication was from the small companies, with a clear majority answering that employment had *not* increased as a consequence.

The larger new companies at Corby (and elsewhere) are highly

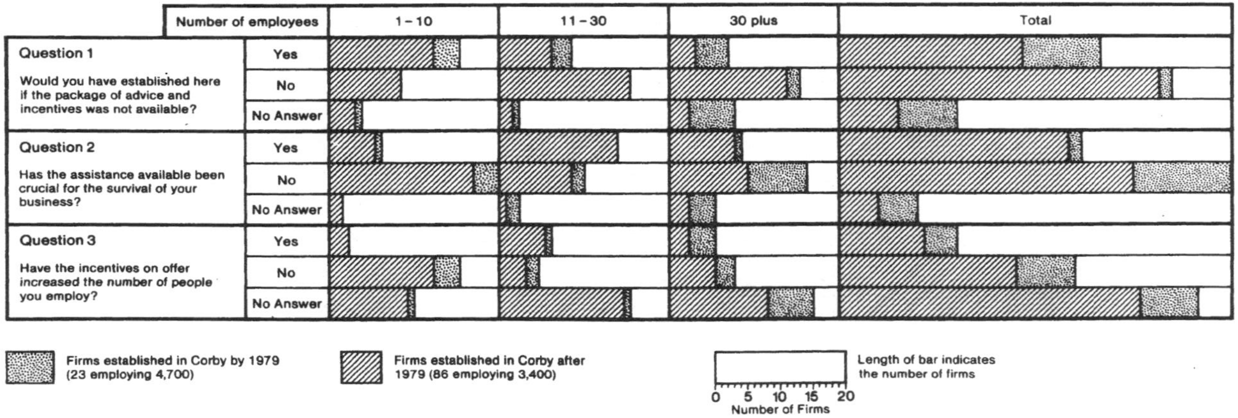

Figure 4.2 Corby survey results

Source: Boulding 1988

vulnerable to international strategic decisions. In 1986 Commodore International, one of Corby's main new 'high-tech' branch plants, announced that it intended to cease all manufacturing operations in the UK. Only 170 headquarters staff would remain at the Corby plant where 1,100 jobs had been promised only two years previously. Explaining this decision company president Thomas Rattigan commented:

> Commodore's major priority is to meet the competitive challenge of the next two to three years, and to do this, the company is going to be increasingly dependent upon fewer and higher technology plants. Corby, being essentially an assembly plant, does not easily fit into this strategy.
>
> (quoted in *Financial Times* 3 January 1986)

And just as in Derwentside and Hartlepool, local critics have complained about the quality of new jobs (see Weston 1985). Labour district councillor Tom McIntosh explained how:

> The council have done all they can to promote the right to work. It's an entrepreneur's paradise. But the council have neglected to emphasize that the social costs to the community in having low-wage, non-unionized factory units can be very demoralizing. From their optimism there's a kind of built-in pessimism. The council have got carried away by their own propaganda.
>
> (interviewed on BBC Radio 4, 4 March 1986)

The re-industrialization of Corby is often presented as a success story; yet its location within the UK is one which confers it considerable advantages over other steel-closure areas. It is interesting to speculate whether the designation of Enterprise Zone status to this particular town was in fact one way of ensuring the partial success of at least one Enterprise Zone. There is clearly also a need to look beneath the figures at the quality of new jobs. A different steel closure area in the UK which has both consciously eschewed Enterprise Zone policy, and adopted an employment strategy which focuses more specifically on qualitative issues, is Sheffield in South Yorkshire.

An alternative strategy: Sheffield

De-industrialization of the local economy has transformed Sheffield from a steel city to a service city. In 1971 half its working population of 281,000 was employed in manufacturing industry; by

1987 the sector provided fewer than 58,000 out of the remaining 225,000 jobs (Sheffield City Council 1987: 7). With the establishment of a Department of Employment in 1981 (later renamed the Department of Employment and Economic Development) the City Council, backed by South Yorkshire County Council until its demise, pursued a range of often innovatory policies. From the outset it recognized that the new department did not have the resources to create jobs on a scale necessary to match the city's employment crisis; rather, its brief was to develop training projects, improve the quality of employment, and promote alternative strategies for employment. These included support for co-operatives, managed workshops, and a product development programme. A stringent contract compliance policy meant that companies were requested or obliged to recognize trade unions, and to provide them with access to information. An employment and environmental plan was developed for the Lower Don Valley area, where steel closures had been concentrated. This explicitly excluded the possibility of Enterprise Zone status. It argued instead that there was a need for a campaign to 'draw people together to challenge the logic of the market', which would 'provide a useful basis to challenge the notion of an Enterprise Zone, by emphasising that unfettered capitalism could hardly be expected to regenerate areas it has already destroyed' (Sheffield City Council 1984b: 16).

In many senses Sheffield was at the forefront in the development of alternative industry and employment policies in the early 1980s. In 1985 its 'Jobs Audit' demonstrated the significance of local authority spending to employment in Sheffield, directly and indirectly supporting almost one-fifth of all jobs in the city (see Clark *et al.* 1986). The effectiveness and efficiency of this spending was praised by the chairman of the Audit Commission, John Banham, in 1985, who concluded that 'the best local government is superb and private enterprise could never improve on it, with Sheffield a shining example' (quoted in Sheffield City Council 1986: 2). And in 1987 an outline employment plan set a target of 25,000 new jobs over two years (Sheffield City Council 1987). With the election of a Conservative government for a third term in June 1987, however, the extent to which the authority would be able to fulfil its objectives was placed in jeopardy.

Echoing the reaction to criticism in places like Consett, local business in Sheffield has frequently been at loggerheads with the city council. Mr John Hambridge, chief executive of the city's Chamber of Trade, commented in 1985 that 'there is a strong heart beating in Sheffield, but if we talk pessimistically too much, we will find ourselves down again' (quoted in *Financial Times* 15 January

1985). Almost exactly one year later he was to comment with more enthusiasm on what he saw as a shift in the council's policies: 'We have seen some big changes in the way the council operates and we are working a lot better with them and with less friction' (quoted in *Financial Times* 21 January 1986).

These changes partly reflected the tide of politics. Sheffield City Council's efforts became increasingly directed more to attempting to prevent designation as one of a new generation of Urban Development Corporations (UDCs), wherein local government's planning powers were effectively usurped by central government-appointed bodies. Even the support of the local business community and a specially commissioned consultants' report recommending that a UDC would be inappropriate failed to sway the argument. In March 1988 a new UDC was announced for the Lower Don Valley with a budget of £50 million over seven years, and sweeping control over land-use. The role of Sheffield City Council in the field of economic regeneration had been drastically curtailed.

These relationships between the private sector and local and national government are central to an evaluation of the effectiveness of re-industrialization policies in steel closure areas. A striking phenomenon in the UK in recent years has been the convergence of ideas from all political parties on the need for local intervention in the economy. There are big differences though, as we have seen, in how this should be organized within the framework of a national policy. This is also clearly apparent in other countries, especially France.

France: the Nord-Pas de Calais region

The Nord-Pas de Calais region of France has suffered considerably from the effects of decline in its traditional industries of coal, steel, textiles, and shipbuilding (see Figure 4.3). Just as in the UK, a range of policy initiatives has been devised at both local and national levels in an attempt to replace these lost jobs. The problems which the 1981–86 Socialist administration encountered in formulating and implementing interventionist policies towards steel production were mirrored in the development of re-industrialization policies. Tensions produced by the transition from an outwardly left-wing programme of expansion to more cautious and restrictive supply-side policies were nowhere more apparent than in the Nord-Pas de Calais, a traditional stronghold of the Socialist and Communist parties.

The powers of local government in France were exceptionally limited until the *Loi Deferre* of 1982. Introduced as part of the

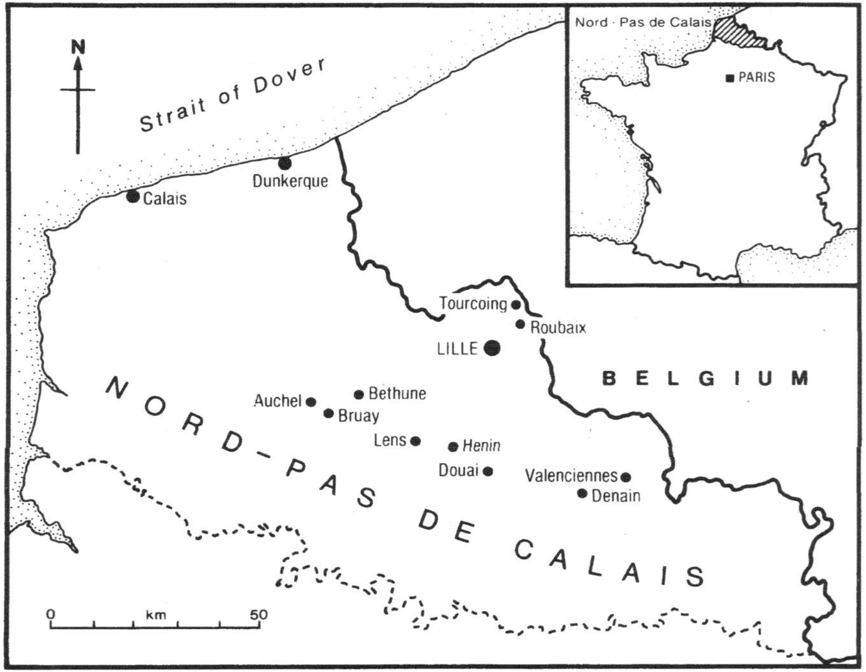

Figure 4.3 Nord-Pas de Calais, France: main towns

post-1981 Socialist government's election commitment to decentralization, this act contained three major changes. Central government control over the acts of local authorities was abolished with checks through the courts replacing administrative and political censorship. Prefects ceased to be regional executives and became instead commissioners responsible for the co-ordination of central government services within the region. And local government could, for the first time, take measures in its own right to ensure the protection of the economic and social interests of its population. In practice this was largely a ratification of policies which the previous government had sought to prevent but had not been able completely to forbid. Pierre Mauroy at Lille, for example, had created several instruments of economic policy which effectively circumvented the constitution's prohibition. With the new law, however, local government was free to devise regional and local economic policy without fear of veto from central government (see also Ashford 1983; Keating and Hainsworth 1986; Mény 1983).

Under these circumstances, the Regional Council of the Nord-Pas de Calais was not slow to extend its range of economic and social policies. An umbrella organization, the Regional Development Agency, was appointed in 1982 and its first operating body, the Economic Development Directorate, in June. This effectively led to a range of Regional Council policies aimed at attracting, financing, and developing new and existing businesses. Direct subsidies to investment took a variety of forms — the *prime d'aménagement du territoire* (PAT); *prime régionale à l'emploi* (PRE); and *prime régionale à la création d'entreprise* (PRCE). The Regional Development Agency encouraged the development of local initiatives, with an estimated 2,000 unemployed people creating their own new small businesses each year in part due to the support and advice offered, and advised existing businesses. It also took a 9 per cent share in the Regional Development Society, a body charged with making medium- and long-term loans to locally established businesses.

After defeats in the municipal elections in 1983 the Socialist party put less emphasis on decentralization reforms. The economic problems of regions like Nord-Pas de Calais remained intransigent, however. Indeed, they were soon made worse by other central government decisions concerning the nationalized coal and steel industries (see Zukin 1985). These were ultimately to be of decisive political significance to the future of the coalition government. In recognition of the unpopularity of policies towards coal and steel, a new range of policy instruments was devised and existing measures were strengthened in an attempt to legitimize the policies of the big nationalized industries. These were to prove both ineffective and insufficient to counter growing hostility to the government from within its own power bases.

With opposition from even its traditional supporters mounting (especially over cut-backs in the coalmining industry) and speculation over the future financing of the steel industry rife, in January 1984 the government announced the outline of a new two-year re-training programme closely tied to support for the creation of new businesses in regions such as Nord-Pas de Calais and Alsace-Lorraine. Between fifteen and forty special zones were to be set up in which investment would be encouraged by a combination of financial incentives and relaxation of financial and social legislation. Regulations to be waived *might* include those under which companies with more than nine employees paid an additional 3 per cent of their salary bill in social security contributions and taxes; or those through which companies with more than eleven workers were obliged to appoint a worker's delegate, and those with more than fifty employees, a works committee. A further announcement of relevance to the threatened industries was Renault's decision to prepare measures to encourage the repatriation of immigrant workers (who made up 17,000 of the company's 102,000 workforce), including a possible doubling of the repatriation allowance to Fr. 40,000. Union reaction to the government's proposals was strong and predictable. Henri Krasucki, the CGT union leader, spoke of 'unbelievable attacks' being made on workers' rights and 'exorbitant privileges' granted to employers. 'The time has come,' he said, 'to sound the alarm for mass trade union action' (quoted in *Financial Times* 31 January 1984). Both the CGT and the CFDT were becoming anxious to distance themselves from government.

More details of the proposed reconversion schemes were announced later. Two main policies were envisaged. Under the first, workers made redundant from the coal, steel, and shipbuilding industries were to be eligible for two re-training schemes on 70 per cent of their former pay. Some 10,000 to 15,000 people were expected to benefit. The second measure envisaged the establishment of some ten 'perimeters of re-birth' in depressed regions, with tax and credit incentives to business and a minimization of regulations. The cost of these zones was estimated at Fr. 5,000–6,000 million. French employers reacted with disappointment and trade unions with scepticism.[2]

In April the implications of the new policy direction for steel were revealed: 25,000 further job losses by 1987 were announced, to the outrage of the Communist coalition partners. Strike action paralysed much of the most severely affected region, Lorraine, as unionists battled with riot police (see Chapter 3). Of ominous significance to Nord-Pas de Calais was an announcement that while a Fr. 500 million aid fund for the Lorraine region (foreshadowed a year previously) would soon be set up, no further measures should be expected. More modest measures such as other nationalized industries being steered into the region (Thomson was already building a plant to make video recorder parts at Longwy) were all that was on offer, and these would remain comparatively small projects.

In July the Communist party left the ruling coalition government in protest against the new tenor of policy and in September the Socialist government announced expanded plans for re-training schemes, as the onus shifted further towards supply-side policies. These were general labour market policies, and were not specific to steel closure areas or ex-steelworkers.[3] More and more the meaning of socialism in France in the 1980s was being redefined as the minority Socialists were following an anti-inflationary, supply-side policy with minimal intervention in the market. Within the Nord-Pas de Calais, opposition to this was mounting.[4] The unions, too, were increasingly distancing themselves from government employment policy. In June 1985 even Edmond Maire of the pro-Socialist CFDT denounced it as a resounding failure.

In July this opposition crystallized around the steel industry, with the unexpected announcement of closure at Trith-Saint-Leger near Valenciennes, with the loss of 770 jobs, and a further 1,400 job cuts elsewhere (see Chapter 3). This struck at the heart of the Socialist party. A shop-steward declared that 'now we know who are the con-men, the murderers of steel; Mitterrand and Fabius. The Socialists here are finished' (quoted in *Financial Times* 15 October 1985). In an attempt to shore up declining support in one of its former heartlands, and in response to pleas from the local Socialist party, the government unveiled a series of measures for the region in October. The programme aimed to create up to 15,000 new jobs. In direct response to the Trith-Saint-Leger closure, Thomson was to build an electronic components plant there, creating 350 new jobs, rising perhaps to 600 in the longer term (though the question of who would get the new jobs, and on what conditions, remained unasked and unanswered). Electricité de France and Air Liquide were also to locate a new plant in the region to produce hydrogen for the Ariane space rocket. In certain areas — Roubaix and Tourcoing, Valenciennes, the Sambre valley and the mining district — businesses creating new jobs would have their social charges reduced by 30 per cent in the first year, 20 per cent in the second, and 10 per cent in the third year. This measure, it was hoped, would help to create up to 10,000 new jobs. Sodinor, the steel company job-creation subsidiary, operating in Valenciennes, Dunkerque, and the Sambre valley (having already created an estimated 3,800 jobs) was given a further Fr. 150 million to create another 5,000. An unspecified amount was also to be added to the regional *prime d'aménagement du territoire* (PAT) which (it was claimed) had created 14,000 jobs since 1982. The *fond régional productique*, launched in 1985 as a 'pump-primer' to encourage modernizing investments in traditional industries, was extended to 1986 with a further Fr. 100 million from the government and Fr. 50 million from the European Community. A new, high-tech university was to be built at Lille to encourage further the adoption of new technology. And finally, a number of infrastructural projects was proposed, including a new autoroute from the Belgian frontier, along the coast to Le Havre; and the advantages of the proposed cross-channel fixed link were to be examined (for details see Région Nord-Pas de Calais 1985).

The 'Fabius Plan' met with mixed reaction in the region as the national Socialist party conference met later in the month. Here the reorientation of the party was completed. One senior member even saw it as 'the end of one party and the beginning of another' (quoted in *Financial Times* 15 October 1985). The programme was by now unashamedly centrist with job flexibility and partial de-nationalization high on the agenda. Contrasts between this stance and the 1981 election platform, and the failure of government efforts to replace jobs lost in traditional industries, were apparent. And in this environment, outright defeat for the government in the March 1986 elections was seemingly inevitable. The poll gave the Socialist party 31 per cent of the vote against 38 per cent in 1986. The Communist party's share of the vote dropped from 16 per cent to 10 per cent.

The reasons for the Socialist party's dramatic decline in popularity since 1981 were in many senses both reflected in and related to the party's policies towards nationalized industries and correspondingly, to the older industrial regions where those industries dominated and party support was once strongest. By 1986 the Socialists were more concerned with privatizing these assets than safeguarding employment in the regions. 'Nationalization was necessary for the survival of our company', the Socialist-appointed chairman of one of the industries taken over in 1982 said before the 1986 election, 'now for our continuing survival, de-nationalization will be favourable. Whoever wins the election, we are likely to be private in a year's time' (quoted in *Financial Times* 13 January 1986). Budgetary pressures and continuing international competition gradually led the Socialist party in the direction of supply-side policies, to the extent that its stance on privatization more closely resembled contemporary Conservative party policy in the UK than its original 1981 election platform. Just as in the UK any sense of or attempt at co-ordination among the publicly owned companies, in a more rational fashion for all concerned, was conspicuous by its absence.

European Community policies towards steel closure areas

Re-industrialization policies have not been confined to national state governments. The European Community has also evolved a series of measures to mitigate the consequences of steel's decline, partly because through its involvement in regulating production, it has felt a need to be seen to be doing something about the consequences of its own policies.[5] Decline in the Community steel industry has had a dramatic impact on the labour force. From 1974 to 1986 Community steel employment fell by one-half as steel companies shed 400,000 jobs (see Figure 4.4). Because of the historical geography of the steel industry, these job losses have been highly locally concentrated in certain towns and regions. Under such conditions the Community has supplemented its conventional policy responses such as the European Regional Development Fund and the European Social Fund with specific measures under the terms of Article 56 of the Treaty of Paris. As originally drafted in 1951, this provided for the financing of new investment projects through low-interest loans or of non-repayable aid to redundant workers only if the introduction 'of new *technical* processes or equipment should lead to an exceptionally large reduction in labour requirements in the coal or the steel industry' (emphasis added). In 1960, however, in the midst of massive job loss in the coal industry due chiefly not to technical change but to competition from other fuels, a second clause was introduced authorizing the payment of loans 'for the creation of new and economically sound activities or for the conversion of existing undertakings capable of re-absorbing the redundant workers' (Article 56.2(a)), and the payment of allowances to workers (Article 56.2(b)) if fundamental changes in *market* conditions for the coal or the steel industry should compel some undertakings permanently to discontinue, curtail, or change their activities.[6]

Loans under Article 56.2(a) for industrial conversion are requested by the government of the member state concerned. The distribution of these loans therefore largely reflects national policies for the redevelopment of coalmining or steel closure areas (see Figure 4.5). From 1981 loans were granted in some instances

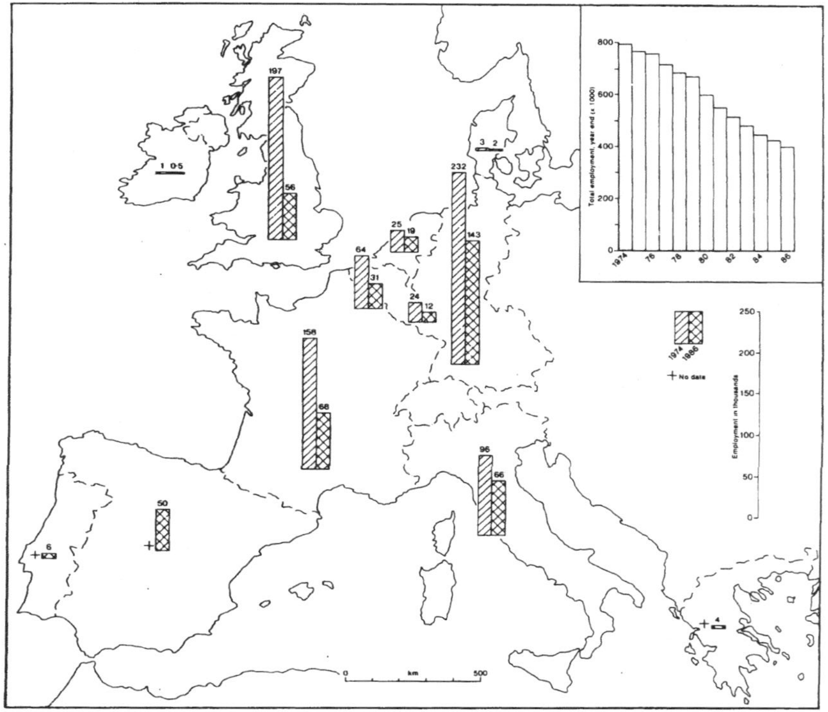

Figure 4.4 European Community steel employment, 1974–86

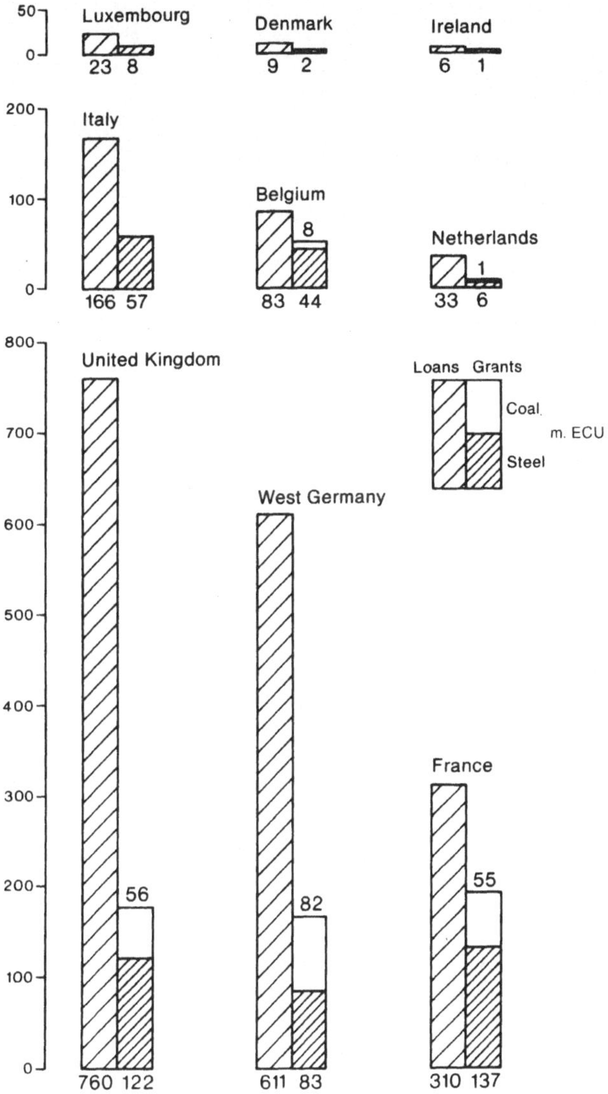

Figure 4.5 Distribution of European Community loans for industrial conversion under Article 56.2(a)

Table 4.1 ECSC loans by sector, 1974–85 (% of total)

	1974	*1975*	*1976*	*1977*	*1978*	*1979*	*1980*	*1981*	*1982*	*1983*	*1984*	*1985*
Coal industry	16.5	20.0	16.9	29.3	32.9	35.2	22.6	7.8	34.6	33.0	1.5	—
Steel industry	66.3	70.5	68.3	66.5	46.5	46.8	41.1	57.7	18.9	28.6	32.6	42.0
Thermal power stations	3.0	—	—	—	3.9	6.1	8.7	6.9	5.5	16.5	8.0	6.0
Industrial conversion	13.1	7.2	5.9	2.2	14.2	8.7	25.8	22.2	22.0	17.9	30.0	35.9
Workers' housing	1.1	2.3	1.7	1.6	2.1	3.2	0.9	3.8	2.4	2.6	4.4	1.7
Iron-ore mines	—	—	7.2	—	0.3	—	—	—	9.7	—	16.8	11.3
Miscellaneous	—	—	—	0.4	0.1	—	0.9	1.6	6.9	1.4	6.7	3.0
Total (%)	100	100	100	100	100	100	100	100	100	100	100	100
Total amount (m ECU)	378	805	1,064	742	797	676	1,031	388	741	778	826	1,011

Source: ECSC *Annual Financial Reports* 1974–86

as 'global loans' to financial institutions, with the intention that they acted as intermediaries for small and medium-sized businesses. Over the period 1980–85 loans for industrial conversion topped 30 per cent of ECSC loans granted (Table 4.1). They had become a significant element in Community activities.

Under Article 56.2(b) payment can be made to redundant workers in the form of tideover allowances, assistance with resettlement costs or help with retraining programmes; and payment can be made to coal and steel companies which continue to pay in full employees on short-time working. The ECSC contribution is subject to the condition that the member state pays at least an equal amount. Payment is within the framework of a series of Bilateral Conventions negotiated between member states and the Community, and is financed by a levy charged annually on each steel producer. The Bilateral Conventions have come to extend the range of aid payable to include lump-sum redundancy payments and finance for early retirement as it has proved increasingly difficult to adhere to the original goal of re-employment (see also Fevre 1987).

With a rapid increase in the number of workers eligible for re-adaptation benefit payments under Article 56.2(b) as more and more companies closed capacity permanently, there was pressure for the Commission to make available extra resources beyond those allocated in the normal ECSC operating budget. The Commission first put forward proposals to the Council in October 1978, calling for Article 95 of the Treaty of Paris to be used as the basis for payment for a series of measures to save jobs through, for example, restrictions on overtime.[7] This was in contradiction to the spirit of Article 56 as applied to that date, where payment was permitted only in circumstances of *permanent* loss of jobs. In the absence of a decision by the Council, the Commission resubmitted its proposals in October 1980, specifying two types of aid: a first for financing early retirement measures, and a second for financing adjustments to the conditions and duration of working hours through, for example, overtime restrictions, an additional shift or short-time working.[8] The legal basis for both types of aid rested in Article 56.2(b) although the second had rarely been applied to that date. The Commission proposal provided for a total of 212 million ECU to be allocated to these measures (termed the *social volet*). It also requested the Council to transfer additional finance from the ECSC general budget to the operating budget since the traditional revenue from a levy on steel producers would be insufficient. The planned sources of finance envisaged transfers of 62 million ECU in 1981 and 50 million in each of 1982 and 1983, along with a special contribution of 50 million ECU by the member states to the

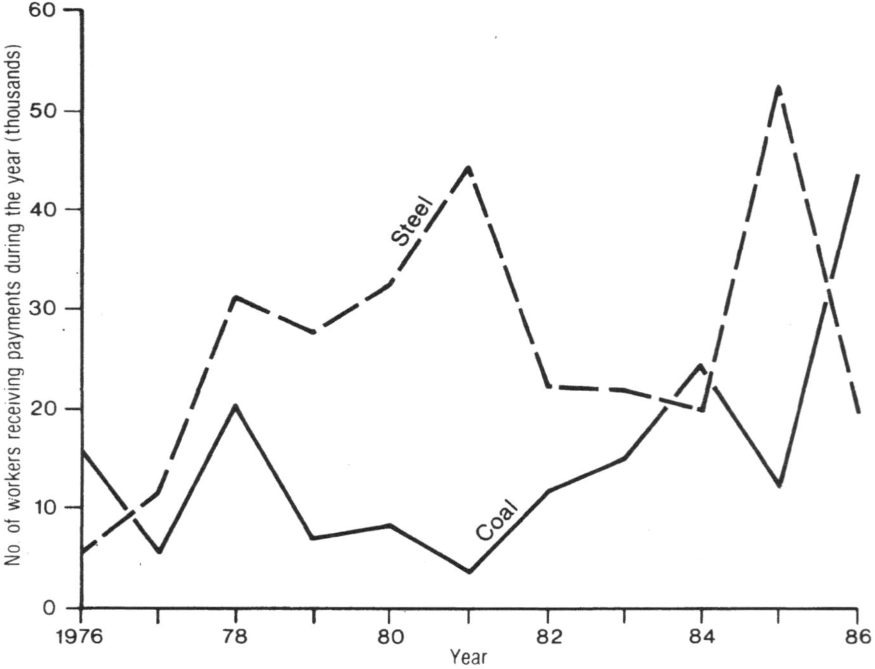

Figure 4.6 European Community payments to steelworkers under Article 56.2(b)

ECSC operational budget in 1981. The Council reluctantly approved these transfers.

With the development of new schemes to support social measures in coal and steel regions, and these industries' continuing problems, payment under Article 56.2(b) assumed great significance in the ECSC budget at a time of growing concern over Community financing generally. The number of steelworkers for whom companies and governments were claiming aid rose to 44,000 in 1981 (see Figure 4.6). The amount made available for payment to steel companies escalated from 4 million ECU in 1976 to 110 million in 1981.[9] Amidst growing concern over the allocation of ECSC finance of this magnitude, the European Court of Auditors examined the system of payments in 1982. It concluded that 'no operational Community objectives have been established and it is therefore impossible to appraise the effectiveness of the Community measures'. Hence, it argued, 'the ECSC aid constitutes no more than a reimbursement of a given amount to the national budgets'.[10] This concern was all the more pressing in that it could not be separated from the politically sensitive question of the distribution of the total Community budget amongst member states.

By the end of 1982 the programme of measures for which additional *social volet* support had been allocated was concluded in all but one country, West Germany. The Commission therefore proposed in April 1983 to extend the social support measures with an additional 330 million ECU for the ECSC operating budget from the general budget over the period 1983–86.[11] During 1984 122.5 million ECU was allocated for such measures in the steel and coal industries but no further transfers were made as the Council refused to extend the exceptional scheme. In 1985 215 million ECU was allocated by the Commission for a three-year programme affecting 67,000 workers, but this was financed from the conventional source for Article 56.2(b) funding (the Commission's operating budget) and did not draw on the general Community budget. As 'social support' measures were phased out the number of workers affected by traditional aid increased once again, reaching 53,000 in steel in 1984 and 44,000 in coal in 1986 (see Figure 4.6), but no additional finance was forthcoming from an already financially strained Community. In a similar fashion to member states, the limits of what the EC could do in response to a problem in which it was integrally involved from the outset had been politically determined by budgetary or fiscal constraints, made all the more acute in the light of pressure from competing member states' interests.

Evaluating re-industrialization policies

Re-industrialization policies are born of a strategy which excludes the possibility of any kind of rational planning within and between public corporations, from governments formed by parties seemingly as diverse as the British Labour and Conservative and the French right-wing and Socialist parties. In part this is a reaction to previous failures in flawed attempts at planning for growth: in the UK in the 1960s and early 1970s, and in France in the early 1980s. It is central also to an economic policy where the state is seemingly left only to pick up the pieces by maintaining a degree of social and political order, although in practice even such a 'market-led' strategy requires a considerable degree of state intervention. Partly a consequence of the effects of earlier government policies towards production, re-industrialization in both national and embryonic supranational states such as the EC has been heavily influenced by financial and budgetary considerations. It is in this context that re-industrialization policies should be evaluated.

We have argued that the scale of the problem is such that there is no way in which the creation of new small businesses can compensate for job losses in traditional industries. It is vital to place this sense of perspective upon such policies. In the words of one commentator, 'the closure of the Consett works in 1980 meant the loss of approximately the same number of jobs as were created in every new independent business establishment in County Durham *over the whole of the 1965–1981 period*' (Storey 1986: 2; emphasis added). When placed against the scale of job losses elsewhere in the economy, local economic policies have neatly been characterized as an attempt to drain the ocean with a teaspoon (Cochrane 1985).

Equally there is considerable uncertainty over the definition of 'new' jobs. The collection of information on this is deeply problematic. In Derwentside, for example, it has led to a number of heated exchanges when attempts were made to elicit such basic data. Yet there is a broad range of opinion which concurs that new jobs should be clearly defined. One authoritative study, for example, which argued in favour of local economic initiatives, adopted several stringent elements in its definition of 'new jobs':

> 1 A new job must be incremental to the stock of existing employment opportunities and must therefore gainfully employ an individual who has not been so employed before he or she takes up this new job.
>
> 2 The adoption of a new job must not mean the loss of a job elsewhere (e.g. by a new shop putting another shop out of business).

> 3 The job should be 'permanent' (of at least twelve months' duration); that is to say, a short-term arrangement which is known in advance to be so would not count as an incremental job.
>
> (Todd 1984: 13)

Against this cry for caution has to be placed a continuing promotional emphasis on the part of those implementing such strategy. To be *seen* to be successful, it seems, is all-important. This is significant on a broader plane too in allowing others to point to the apparent success of policies in action. Derwentside has frequently been used in this respect, often at politically significant moments. For example in March 1987 (on the eve of a UK general election), Industry Minister Giles Shaw commented that Consett's 'recovery' was a

> remarkable success story for a town faced with problems that were traumatic in the past. What Consett has to offer is a manufacturing facility for the UK which is very attractive and a workforce that has shown itself to be adaptable and of high quality.
>
> (quoted in *Newcastle Journal* 31 March 1987)

The 'success' of places like Consett and Corby has been used to demonstrate the appropriateness of such measures elsewhere. When Ian MacGregor launched the NCB (Enterprise) scheme in 1985 he specifically pointed to DIDA's success in revitalizing Consett as a model for the coalfields to follow. Similarly the Enterprise Zone at Corby was singled out in a survey monitoring the initiative nationally as the most successful on almost all criteria (Roger Tym and Partners 1984). We have seen how the term 'success' can be used only in a relative sense. However, this 'success' is in danger of being self-defeating. One consequence of pointing to Consett and Corby as the model for this type of policy response has been the creation of a plethora of similar agencies across the UK. Within the north-east for example, Consett now faces intense competition from South Tyneside, North Durham, Easington and Shildon with their own enterprise agencies, and from Middlesborough, Hartlepool, and Gateshead with their own Enterprise Zones. The paradox is that despite the evident lack of success in Consett, the creation of similar agencies elsewhere on the basis of Consett's claimed success can have the effect only of increasing the problems for places such as Consett and Corby.

It is frequently argued that in Britain practically the only economic policy option now available to local government is via such a public/private sector route. Local politicians admit that they know it is not a satisfactory solution, but ask what else they can do, especially as their constituents expect them to do *something*. Budgets for local authority economic initiatives are tightly constrained by central government direction. Under these circumstances local politicians feel that they have no choice, no option, but to promote *their* place, the image of their town or district, if necessary at the expense of somebody else's place. This territorial chauvinism is by no means a new feature of capitalist society. What is new, and in many respects disturbing, is the extent to which this rivalry is now seen as the only (at that unsatisfactory) answer to unemployment problems of a long-term nature within society. Effectively the orthodoxy of the market-place espoused by central government has found expression also in a place-market (see Robinson and Sadler 1985). What is in question is the share of the cake rather than the definition of its size and type.

It is important therefore not to lose sight of the political significance of organizations like BSC(I). On its formation by BSC, it was plainly seen as a means of smoothing the corporation's investment and redundancy programme, of 'buying off' opposition to closures. A main board paper commented in 1975 that BSC(I) should be viewed as 'an integral element of the Corporation's investment for the future because, if it is necessary, it is implicit that without this element, the reinvestment programme will not produce the results it is designed to achieve'.[12] Not that BSC(I) was concerned specifically about ex-steelworkers. As Roger Thackery, chief executive of BSC(I), was later to explain:

> We have never tried to persuade former steelworkers to set up in business — in some cases we have dissuaded them from it. Anyone employed for twenty or thirty years in a large company is not going to have the skills which make an entrepreneur. What we are engaged in is diversifying the economic base of local communities which had relied on one industry.
>
> (quoted in *Financial Times* 2 June 1987)

Such views were clearly close to the government of the day, especially in the significance attached to the development of a flexible labour force in the UK:

> Basic to the BSC Industry initiative is the understanding that patterns of work and job opportunities are already beginning to change radically. School-leavers and young adults will have to develop a more flexible attitude to work and how to gain a

> livelihood. . . . Changes to long-held beliefs about the nature of work will not be easy to achieve but the future of employment and an important part of the UK's industrial future will depend on their taking place.
>
> (BSC *Steel News*, Job Creation Supplement, March 1987)

In places like Derwentside the real meaning of 'flexibility' is only too apparent. And it is that, as much as anything, which has led to a considerable questioning of the practicality and ethics of the conventional re-industrialization strategy. The cornerstone of the Derwentside Unemployed Group's thinking is that there will be no return to conventional full employment. A member of the group, John Kearney, commented in 1984 that:

> To a great extent Consett presents itself as some sort of microcosm. The industrial ethic has been demonstrated to be totally irrelevant — we find ourselves facing that before the rest of the region and possibly before western Europe.
>
> (quoted in *Sunday Times* 10 June 1984)

Sentiments echoed by Durham County and Derwentside District Councillor, David Hodgson:

> We've got to stop pretending that unemployment will go away tomorrow. Industrial strategies simply set one area competing against another, put one group of unemployed people fighting against another. This is a long-term problem which needs a long-term solution.
>
> (quoted in *Newcastle Journal* 28 May 1985)

And yet, as Kearney recognized at a public meeting in Durham in 1986, such comments pose uncomfortable questions not only for the Conservative government, but also for the Labour party (a political dilemma which was only too apparent in France during the 1981–86 Socialist administration):

> Worklessness is a fundamental crisis for the left and for the right, and any kind of politics which doesn't accept this is a sham. What we're facing is a crisis of political organisation, where the structures which we have are no longer capable of solving our problems for us. Labourism is currently about bigger bribes for industrialists to relocate. The price of maintaining Labourism is to further victimise the victims.

There clearly is a need for local (and national) intervention in the economy, but what has yet to be formulated in terms of a viable politics which does not 'blame the victims' is a conception of how this would relate to national policies and priorities and how aims and objectives could be articulated in participation with the people living in these places. Clearly the 1947 Morrisonian model of nationalized industries and its equivalent in countries like France in practice became a mechanism for the destruction of key industries and communities, not least because it left no room for the needs of working-class communities in the old industrial localities to be met via centralized decision-making procedures. An alternative for the 1990s has to find a way to relate national policies for key industrial sectors with the legitimate aspirations of working people to work, live, and learn in their places — and this, as experience in Sheffield has shown, is crucially dependent upon political will at *both* local and national levels.

Notes

1 In the UK it was later mirrored by British Coal (Enterprise) Ltd, established during the 1984–85 miners' strike and, for a short period, British Shipbuilders (Enterprise) Ltd, created in 1986 but wound up the following year. Similar schemes were also funded by British Rail (Engineering) Ltd and in France by the nationalized coal and steel industries (see Todd 1984).

2 Charbonnages de France (CdF) announced in March that 30,000 of the coal industry's 57,000 jobs were to be shed inside five years. Annual coal output was to be cut from 18 m tonnes to 11–13 m tonnes. The break with election promises for the industry was complete, for in 1981 the Socialists had promised to boost annual output to 30 m tonnes by 1990. Government subsidy to the industry was to be frozen at a constant Fr. 6,500 million a year for the next five years, a substantial cut in real terms. At the same time the industry was to receive Fr. 325 million a year to help re-industrialization of the coalmining regions. Union reaction was predictable: 5,000 miners demonstrated in Paris and a national one-day strike was called in the industry.

3 There were two elements to these plans. Under the first, 750,000 training places for people aged 16–25 were to be created, largely in state education but with up to 300,000 in private industry. By agreement between the government and the employers, this three-stage professional training scheme would last for up to three and a half years before the trainee even qualified for a permanent job. The second part of the package involved a programme of *Travaux d'utilité collective* whereby local government and major national organizations would take on unemployed young people for work of benefit to the community. The central government offered Fr. 1,200 a month for this, while the local authority could (but had no

obligation to) add a further Fr. 500 a month. This was an extension of a previous scheme which had created only 15,000 of a hoped-for 210,000 places. Trade unions were extremely cautious of its potential effect in restricting regular full-time public sector recruitment.

4 In December 1984 the Regional Council passed a motion condemning the delay in the introduction of new redevelopment programmes and calling for guarantees for the continued future of coalmining within the region. It deeply regretted that other essential measures to safeguard the regional economy were not being put into effect (see Région Nord-Pas de Calais 1984).

5 The perceived need for such measures was encapsulated by the European Commissioner responsible for installing the steel quota regime:

> I would like to emphasise here that there is one idea that I cannot possibly accept: the idea that, in the Community, it is Brussels which is responsible for the reorganisation, in other words the negative, unpopular and difficult side, while it is the national governments which are responsible for improvements and conversions, in other words the positive aspects: loans, social welfare, aid, etc. This is an unacceptable division of responsibility.
>
> (Davignon 1980: 508)

6 *Official Journal of the European Communities* 33, 16 May 1960, p. 781.

7 Com (78) 570, 31 October 1978, revised as Com (79) 436, 20 July 1979.

8 Com (80) 676, 28 October 1980.

9 Commission of the European Communities, *General Reports on the Activities of the European Communities*, 1976–86.

10 *Report by the Court of Auditors on the accounting and financial management* (Annex to the Annual Report ECSC, 1982) Luxembourg, 21 December 1983, pp. 23–4.

11 Com (83) 158, 13 April 1983.

12 *British Steel Corporation: New Jobs — Augmenting Government Efforts to Create New Employment in Closure Areas*, Memorandum for Board meeting, 31 July 1975.

Chapter five

Future directions for steel towns, steelworkers, and the steel industry

Introduction

The formulation of state policy towards steel closure areas exemplifies how the relations between local and national politics and policy are currently being redefined in the UK. An outwardly simple populist ideology backed by an upsurge of carefully orchestrated nationalist sentiment and economic policies which tend to favour the right voters in the right place at the right time have combined with a collapse in disarray of formal political alternatives (symbolized by resort to the International Monetary Fund in 1976) to produce a seemingly stable hegemony, coloured by a 'new realism' as to what is possible in terms of state economic management. This stability, though, is based upon conflict; for it has been established around an appreciation of two major sources of power and an appropriation of policies towards them. The UK state is currently characterized by a strong centralizing tendency and a powerful privatizing process. The latter is sold as a means of economic emancipation; the former as an adjunct, an unfortunate necessity because of the policies of some previous town hall inhabitants. This order of policies should not be surprising; for the rhetoric of privatization belies the reality of centralization.

It is important to remember that such changes are by no means confined to the UK and that these are contingent events, only one of a set of possible outcomes from the context provided by a variety of diverse national and, especially, international processes. The current rightist hegemony is born out of a belated recognition of the deep-seated international character of crisis facing much of the UK's manufacturing industrial base and of the failures of previous social democratic policies towards manufacturing, typified by but not confined to steel. Any alternative policy for what remains of the UK's steel industry, for towns and regions blighted by steel's collapse, and for other industries and associated local and regional economies facing similar changes, has to take this centrally into account. Its implications are profound. Such an integrative task is our objective in this concluding chapter. In it we seek to consider the implications of several interconnected issues. We first look at the future prospects for ex-steel towns. Then we consider those places where the steel industry still, for the moment, remains, through an assessment of likely future developments in the global steel industry. We conclude by outlining some of the main political and theoretical lessons to be identified from these trends.

What kind of future for the ex-steel towns?

Gravitation of political power towards the centre is an invidious and deep-rooted process with many manifestations. Standing on the steps of Sheffield's Town Hall on 1 May 1987 (a day not without its own symbolism), one of Sheffield City Council's elected members, Councillor Jim Jamison, made a significant, unsolicited comment. As he looked back at the stone-built offices behind him, an edifice to a legacy of well-managed local government, he reflected on the UK general election then just over a month away. 'What worries me', he said, 'is that if the Tories get a third term of office, there won't be any of this left. There'll be an agency in its place'. The subsequent creation of a non-elected Urban Development Corporation brings home the deep and real foundation of his fears. Meanwhile in other steel closure areas of the UK such as Consett and Corby, elected local governments of various political complexions have eagerly embraced the enterprise agency concept with its similar subversion of economic power to the hands of a privileged few. But how has this come about? How is it that such outwardly radical movements, seemingly charged with the potential for change, exemplified by anti-closure campaigns and alternative local economic strategies, have become so easily pushed aside?

Part of the answer, of course, lies with the fact that in many senses such campaigns and strategies were by no means as radical as first appearances might have suggested. Many anti-closure campaigns became so sadly embroiled in internecine trade union disputes as to lose sight of their original objective. This is not so much a comment on the inherently accommodative nature of many national trade union officials (although this should not be denied; most uphold a strong desire to preserve some form of status quo). It is more a question of the problematic union policy stance taken in the 1960s and 1970s towards capacity expansion and technological change, coupled with an unwillingness to come to terms with some of the consequences of this until it was almost too late. Equally most if not all alternative local economic strategies were, almost by definition, accepting some key external parameters which fatally undermined their likely effectiveness, laying their proponents open to frequent and often painful charges of 'radical managerialism'.

But it cannot be denied that in their sheer existence such campaigns and strategies possessed a degree of potential for change. The implications of events that could bring thousands of people out into the streets and persuade them to attend meetings and rallies should not be under-estimated. They demonstrate, if nothing else, that the issues involved had a powerful degree of personal significance for many people. In the extent to which such action was reported, debated, and analysed, it also had a national and international importance. The physical presence of such demonstrations clearly signified a scarcely veiled, latent challenge to the power of the state and its capacity to govern.

This is not to argue, though, that they represented a revolutionary or radical challenge to the basic social relations of capitalist production. The terrain of debate was over the presence or absence of waged labour, not the waged characteristic of labour in capitalist society. The labour relation was never seriously under threat; rather the probability of unemployment was the catalyst for protest. This distinction between what anti-closure campaigns were about, and what they did not by and large encompass, is important. Not least because it points to the fact that they were clearly structured by generalized acceptance of a number of determining rules of capitalist society; and as such, their likely effectiveness was conditioned by a number of issues that reflected the impacts of circumstance and strategy.

The particular conditions under which campaigns evolved were influential in shaping their likely outcome. In France in the late 1970s the political map was such that outwardly local and regional issues became articulated nationally and were partly instrumental in altering that political balance. By contrast in the UK at almost exactly the same time, the pendulum of power had begun to swing in a very different direction. There, protests fell on increasingly deaf ears.

The question of strategy is also relevant. A whole range of options proved to be open to national and embryonic supranational states effectively to buy off opposition to closures, and in due course to ease confrontation over regeneration policies too. The former included relatively easy policies to sell (literally) such as early retirement, re-training, and earnings compensation; although their financing often proved more problematic. There were, however, also policy areas in which it was more difficult to deliver, notably the promise of alternative employment. Partly as a consequence, this was rapidly devolved to a more local level of intervention in the economy. In part also this was because local politicians were eager to grasp the mettle, sometimes without realizing the full implications of doing so. Once installed, local economic policy mechanisms such as enterprise agencies, backed by a facilitative national policy direction, developed a strongly self-fulfilling mystique of secrecy, which is awkward to penetrate, and an aura of success which is difficult to substantiate.

So what, then, are the likely future prospects for current (and prospective) ex-steel towns and regions? It should be apparent by now that the answer to such a question depends upon a series of decisions. As things stand the likely future for places such as Consett or Longwy is more of the same: an increasing proportion of low-paid, part-time employment, often taken by women; high male unemployment; outmigration; poverty; and continuing state programmes to manage the surplus population. Such a scenario is the result of political decisions which have already been taken and assumes their continuation. It is not inevitable; nor are the policies irreversible, although the longer they are implemented, the more difficult it becomes to reverse their deleterious effects. Before we go on, though, to consider alternatives, we need to look also at another constraint on policy options: the international dimension of production change in the steel industry.

Future prospects for the global steel industry

We have seen how the processes of restructuring which have devastated particular places and regions are ultimately international in scope. The overall decline of steel output and capacity in established producer countries of western Europe and north America, counterposed against expansion in some newly industrializing countries (NICs), is often presented as an inevitable switch associated with steel's status as a somehow 'mature' industry. But we would challenge the notion that steel is in this sense 'mature'. In fact, the evidence points to quite a contrary conclusion. Steel is in many ways a still evolving sector of the economy. It is far from employing a stagnant technology, for example. Significant new developments are underway or in the process of adoption: continuous casting of thinner and thinner slabs, direct injection of coal to blast furnaces, direct reduction of iron ore, improved grades of steel, new steel coatings. The observed pattern of locational change is not predetermined, but rather the latest act in an unfolding drama of continuing global reorganization, reflecting the latest round of corporate decisions and government policies towards trade and production. There is absolutely nothing inevitable about such things. The contrast between the expansionist path adopted in South Korea and Brazil, and that of decline sponsored by UK and French governments since the mid-1970s, should confirm that. We should be wary, therefore, of extrapolating future trends from present developments. But this is not to say that there is no value in examining some of the likely possibilities for the future.

Our central concern has been with the steel industries of the established producer countries. Companies in these countries are clearly aware of the great pressure being exerted upon steel by competitive alternative products, and of the need to develop better qualities of and wider uses for steel. BSC, for instance, has devoted substantial marketing efforts, very successfully, to improve steel's share of building materials used in multi-storey construction. But competition is especially fierce in the automotive industry, perhaps *the* crucial market for steels. United Engineering Steels, one of the UK's major suppliers to this sector, is working with BSC's Swinden Laboratories to develop new kinds of steel, and with GKN's research offshoot, GKN Technology, to devise components which can make better use of that steel.[1]

Despite this substantial commitment to material and process development on the part of some steel manufacturers, the competition from alternative products is becoming increasingly severe, as other materials achieve commercial viability. A detailed report from the Economist Intelligence Unit (1988) highlights just some of these new areas. Porsche has become one of the first vehicle manufacturers to use ceramic parts in an engine. GKN is working on a large transverse spring in composite plastic. Alcan has developed new body structures made of bonded aluminium, claiming many times the rigidity and great weight saving against steel.

Partly in consequence, many one-time steel companies are making great strides to diversify out of dependence upon steel. GKN has been transformed from a steel company into an automotive components producer committed to producing the components, not necessarily to using steel.[2] The major Japanese steel companies are all aggressively expanding their interests in the electronics sector. In a slightly differing strategy the Dutch steel company Hooghovens purchased Kaiser Aluminium Europe Inc. in 1987 (one of the four leading west European rolled and extruded aluminium product manufacturers). As its board of managing directors noted, this acquisition was 'from the strategic point of view, our most important investment in 1987', making aluminium 'a second main activity'. Together with other purchases of downstream, higher value-added steel processing activities, this led to speculation that 1988 might well be 'the first year in Hooghovens' history in which the Group derives more than 50 per cent of its turnover from activities other than steelmaking'.[3] Many of the major West German steel companies have long since reached such a position.

As many long-established steel companies diversify out of steel, continuing question marks will hang over the future of their steelworks. In western Europe and north America, the threat of further closures is likely to be present for some while yet. Early in 1988, for example, the closure of Krupp's massive Rheinhausen works was formally agreed, despite some strong if sporadic protest. A plant which once symbolized post-war West German reconstruction had now become surplus to this company's strategy — and there is much more excess capacity left within the European Community. In the UK an overwhelming body of expert opinion anticipates drastic changes at BSC after privatization. Industry expert Nick Garnett, writing in the *Financial Times* (10 June 1988), had this to say:

> The plant configuration will almost certainly change and shrink over the medium term. The survival of the strip mill at Ravenscraig in Scotland is guaranteed only until next year. The rest of the Ravenscraig plant has a guarantee for another six years or so 'subject to market conditions'. It seems unlikely that Ravenscraig will survive anywhere near that long. . . . In the long term, the successor company to the corporation will probably work towards reducing the five integrated plants to two or three.

And on many accounts BSC will not be alone in this further round of closures, cut-backs, and redundancies.

The problems this situation will pose for established forms of trade unionism in the developed countries will probably be exacerbated by the continuing expansion of steel capacity in other countries. Japanese steel output wobbled heavily in 1986/87 but 1988 saw a return to profits amongst the major companies and talk that earlier planned cut-backs had been excessively pessimistic. Growth in south-east Asia other than Japan has become the most intriguing part of the world production and trade equation. From the perspective of international development, the possibility of such growth should not be denied; nor should the problems of workforce exploitation and the difficulties of labour organization be ignored. The International Metalworkers' Federation (IMF) noted that Asia has become the stronghold of the world steel industry, with a steel workforce of over 1 million, representing two-fifths of steel employment in the non-centrally planned economies.[4] Much of this is non-unionized and in many countries (such as South Korea) great pressures are exerted to discourage any form of trade union activity. Yet even in this apparent economic boom area, the IMF noted the rise of over-capacity and an emerging trend to raise labour productivity and shed jobs, partly in consequence of trade restrictions and partly because of domestic economic difficulties.

To return to our original remarks at the start of this section, then: the future of steel employment in all parts of the world, especially the developed countries, is one of uncertainty. The most probable scenario is one of further destruction of capacity through yet more place-specific devalorization of capital. Put simply, further closures, if not inevitable, appear likely in the advanced capitalist economies; and as more and more developing countries bring on-stream new plant to meet export targets and those of domestic industrialization programmes, this capacity is likely to expand well ahead of domestic and world demand. In a global climate of trade protectionism and recession, this double imbalance — within the NICs on the one hand, and between the NICs and the developed countries on the other — might well produce a potentially explosive global cocktail of closures and cut-backs. At the very least it will be a difficult problem to manage if not resolve; for although the world's major (that is western and Japanese) banks might slow credit for expansion, their current demands for debt-servicing payments only act further to pressurize NIC governments into making steel for sale abroad. The uncertainty of the future should be tinged with apprehension.

Political and theoretical implications

What, then, does all this mean for our political and theoretical considerations? What are the longer-term connotations of these remarks? And how might they inform future political and theoretical developments?

It should be apparent by now, if nothing else, that places are constructed politically. Present, past, and future steel towns and ex-steel towns alike were and are being shaped and fashioned by the actions of capital, labour, and the state. And these actions are clearly political in character: not party political, but in the sense of sharing the capitalist work ethic. What has sharply characterized the last few years, and made the politics of capitalism more openly problematic in several advanced countries, is that as many steel towns have de-industrialized, with the state often a prominent social actor in the process, most steel companies have internationalized and diversified. In this way the fortunes of Consett were and are inextricably bound up with the fate of other communities in the UK and across the globe. In this context the 'locality' has come to assume a great theoretical and political importance; and as a locus for much (but crucially by no means all) of what has happened, it clearly has significance. This should not allow it, though, to become the only social and spatial scale of reference; for localities such as Consett have been and are being constructed, politically, in and by a truly global capitalist economy, within the confines of capitalist property relations and the requirement to produce profitably, mediated by national (and sometimes supranational) states.

Conventional distinctions between the 'economic' and the 'political' can often mislead. One currently dominant strand of bourgeois political concern is 'to let the market decide'. Withdrawal of the European Community from regulation of the steel industry in 1988, with the abandonment in disarray of its quota system, and the privatization of BSC with disarray in this instance confined to the Labour party, are from such a perspective, two clear examples of this tendency in action.[5] But it is crucial to any alternative theorization and strategy to recognize that such an ostensibly market-led regime is a matter of political choice; and, as such, in no way represents the workings of some mythical free and fair market, let alone the expression of some natural process of development. Rather capitalist relations of production and the markets through which they operate are socially constructed political choices. This recognition is often lacking from accounts of the internationalization of production, for national states do not inevitably have to embrace the world market. It is, at the same time, central to the formulation of any truly alternative strategy for industries such as steel.

The recent experiences of alternate right- and left-wing governments in France and the UK should demonstrate the limits to any national attempts at planning for key sectors without regard for the disciplines that the law of value imposes on capitalist production and for local dimensions on the one hand and an international dimension on the other. Clearly a really alternative strategy needs to begin to articulate local and national considerations within a recognizably global context, incorporating too the right of all nations, developing and developed alike, to a steel industry. It needs to come to terms with capital's changing use of place and seek to transcend it. Ultimately this implies challenging the legitimacy of capitalist relations of production and the imperatives that govern production arising from these, to begin to ask what production is *for*, although it is important to stress that there is no easy readily available 'off the shelf' socialist alternative. Rather that is something that will need to emerge from political debate and struggle. But if it is to emerge it needs to begin to do this soon, otherwise the recent experience of steel will be far from uncommon. Many other industries are on the verge of similar changes. Coal and shipbuilding workers already understand this, motor vehicle production is well on the way, and other 'modern' sectors like chemicals lie just round the corner. The increased mobility of capital led to the increased redundancy of places at just the same time as social science rediscovered place. If we want to make people and their places matter again, we have to look carefully at what created them — capitalism — and seek constructively to change it, if not transcend it.

Notes

1 The logic behind these moves has been expressed as follows:

> Development programmes at Swinden Laboratories are aimed at identifying and perfecting improved steels which include new generations of engineering steels with improved machinability — and hence lower machinery costs — and air hardening forging steels which enable component properties to be achieved by air cooling direct from the forge. . . . These developments can result in steels with better consistency and closer control of mechanical properties, less distortion and an improved fatigue life.
>
> At GKN Technology, finite element analysis and CAD systems are employed to maximise the benefit of these steels in component manufacture. By working to an optimum design concept involving design detail and material specifications, lighter and more efficient components can be achieved.
>
> (United Engineering Steels, *Annual Report* 1987: 5)

2 These changes were described by GKN chairman Sir Trevor Holdsworth:

> To crystalise the transformation which has taken place in the composition of the Group's business over the last two decades, we think it appropriate and timely to propose a change in the Company name to 'GKN plc'. The names 'Guest', 'Keen' and 'Nettlefolds' are historically linked with businesses in steel, bolts and nuts and fasteners which have all now ceased to be part of our mainstream strategy; the investments which we retain in steel, whilst important, are in the form of participation in joint ventures. Our principal business is now the development and manufacture of component systems and products for the world's vehicle industries.
>
> (GKN, *Annual Report* 1985)

3 Hooghovens *Annual Report* 1987, p. 11.
4 IMF remarks at the meeting with the OECD Steel Committee, 20 November 1985, Paris.
5 This disarray was clearly in evidence during the second reading in the House of Commons of the British Steel Bill, paving the way for privatization of BSC. Seizing the opportunity, Trade and Industry Minister Kenneth Clarke announced that in his view:

> the nationalisation-denationalisation debate is effectively over in this country. . . . After 40 years the public have seen that the great state corporations dreamt of by Herbert Morrison are not in practice efficient providers of goods and services.

He went on to tease the Opposition benches:

> Steel debates have aroused great passions in the past. . . . Steel is one of the great issues that has caused divisions between parties and within parties — and definitely within the Labour Party at certain times in the past. I do not think that those old battles will be fought again. . . . I shall be surprised if any Hon. Member speaking for the Labour Party threatens to return the Corporation to the public sector, if ever there is a Labour Government.
>
> (*Hansard* 23 February 1988: vol. 1439, col. 170)

Glossary

basic oxygen steel method for the production of steel from iron involving blast of heated oxygen — sometimes referred to as Kaldo or LD process

blast furnace used for the production of iron from iron ore

continuous casting process of drawing steel off from converter in continuous cast of semi-finished product

electric arc alternative route for production of (usually) low volumes of special steel by re-melting scrap steel

flats wide rolled products, including plate (heavy gauge), strip (narrow width, light gauge) and coiled plate

hearth diameter measure of the size of a blast furnace

ingot cast of steel from converter requiring substantial re-rolling — now largely obsolete, replaced by continuous casting

long products narrow gauge rolled products including profiles, sections, beams, and bars

open hearth now obsolete method for the production of steel from iron

semis semi-finished steel, either rolled from ingot or continuously cast — three main shapes referred to as billets, blooms, and slabs

sinter fused amalgam of coke, iron ore, and limestone — raw material in the production of iron in the blast furnace

special steel high value-added grades of steel with particular specifications

torpedo vessel for the transport of hot iron from blast furnace to steel converter by rail

Bibliography

Ardagh, J. (1982) *France in the 1980s*, Harmondsworth: Penguin.

Ashford, D. E. (1983) 'The Socialist reorganisation of French local government — another Jacobin reform?', *Government and Policy* 1: 29–44.

Baker, C. (1982) *Redundancy, Restructuring and the Response of the Labour Movement: A Case Study of British Steel at Corby*, W-P 26, School of Advanced Urban Studies, University of Bristol.

Bassett, P. (1986) *Strike Free: New Industrial Relations in Britain*, London: Macmillan.

Beynon, H. (ed.) (1985) *Digging Deeper: Issues in the Miners' Strike*, London: Verso.

Beynon, H., Hudson, R., and Sadler, D. (1986) 'Nationalised industry policies and the destruction of communities: some evidence from north-east England', *Capital and Class* 29: 27–57.

Bierich, M., Herrhausen, A., and Vogelsang, G. (1983) *Stahlgespräche: Bericht der Moderatoren*, Düsseldorf.

BISF (British Iron and Steel Federation) (1966) *The Steel Industry: Stage I Report of the Development Co-ordinating Committee*, Benson Report, London.

Bleitrach, D. and Chenu, A. (1982) 'Regional planning: regulation or deepening of social contradictions?', in R. Hudson and J. Lewis (eds) *Regional Planning in Europe*, London: Pion.

Bömer, H. (1982) *Neue Beweglichkeit — neue impulse?*, Frankfurt.

Boulding, P. (1988) 'Re-industrialization strategies in steel closure areas in the UK', unpublished Ph.D. thesis, Durham University.

Boulding, P., Hudson, R., and Sadler, D. (1988) 'Consett and Corby: what kind of new era?', *Public Administration Quarterly* 12, 2: 235–55.

Bowen, P. (1976) *Social Control in Industrial Organisations: A Strategic and Occupational Study of British Steelmaking*, London: Routledge.

Bradbury, J. (1982) 'Some geographical implications of the restructuring of the iron ore industry', *Tijdschrift* 73: 295–306.

——— (1985) 'Restructuring in the steel industry: a North American case study', unpublished mimeo.

Bryer, R. A., Brignall, T. J., and Maunders, A. R. (1982) *Accounting for British Steel*, Aldershot: Gower.

Bush, R., Cliffe, L., and Sketchley, P. (1983) 'Steel: the South African connection', *Capital and Class* 20: 65–87.

Buss, T. F. and Redburn, F. S. (1983) *Shutdown at Youngstown: Public Policy for Mass Unemployment*, Albany, NY: State University of New York Press.

Bytheway, B. (1987) 'Redundancy and the older worker', in R. Lee (ed.) *Redundancy, Layoffs and Plant Closures*, Beckenham: Croom Helm.

Carney, J. (1976) 'Capital accumulation and uneven development in Europe: notes on migrant labour', *Antipode* 8: 30–8.

Carney, J. and Hudson, R. (1978) 'Capital, politics and ideology: the North East of England, 1870–1946', *Antipode* 10: 64–78.

Clark, G. L. (1987a) 'Corporate restructuring in the steel industry: adjustment strategies and local labour relations', in G. Sternlieb and J. Hughes (eds) *America's New Economic Geography*, Rutgers University, New Jersey.

—— (1987b) 'Enhancing competitiveness in the US steel industry: the steelworkers' and National Steel's cooperative partnership', unpublished mimeo.

Clark, N., Critchley, R., Hall, D., Kline, R., and Whitfield, D. (1986) 'The Sheffield Council Jobs Audit — why and how?', *Local Economy* 1, 4: 3–21.

Cleveland County Council (1983) *The Economic and Social Importance of the British Steel Corporation to Cleveland*, Middlesbrough.

Cochrane, A. (1983) 'Local economic policies: trying to drain an ocean with a teaspoon', in J. Anderson, S. Duncan, and R. Hudson (eds) *Redundant Spaces in Cities and Regions?*, London: Academic Press.

Coffield, F., Borrill, C., and Marshall, S. (1986) *Growing Up at the Margins: Young Adults in the North East*, Milton Keynes: Open University Press.

Davignon, E. (1980) 'The future of the European steel industry', *Annals of Public and Cooperative Economy* 51: 507–19.

Deitch, C. and Erickson, R. (1987) ' "Save Dorothy": a political response to structural change in the steel industry', in R. Lee (ed.) *Redundancy, Layoffs and Plant Closures*, Beckenham: Croom Helm.

Department of the Environment (1985) *Enterprise Zone Information 1984–85*, London: HMSO.

Department of Trade and Industry (1973) *British Steel Corporation: Ten Year Development Strategy*, Cmnd 5226, London: HMSO.

—— (1978) *British Steel Corporation: The Road to Viability*, Cmnd 7149, London: HMSO.

Dominick, M. (1984) 'Counteracting state aids to steel: a case for international consensus', *Common Market Law Review* 21: 355–404.

Donaldson, M. (1981) 'Steel into the '80s — the rise and rise of BHP', *Journal of Australian Political Economy* 10: 37–45.

Donaldson, M. and Donaldson, T. (1983) 'The crisis in the steel industry', *Journal of Australian Political Economy* 14: 33–43.

Dunford, M. (1988) *Capital, the State and Regional Development*, London: Pion.

Durand, C. (1981) *Chomage et violence: Longwy en lutte*, Paris: Galilée.
Durand, C. and Kourchid, O. (1982) 'Débat sur Longwy en lutte', *Sociologie du Travail* 24: 85–94.
Fevre, R. (1986) 'Contract work in the recession', in H. Purcell *et al.* (eds) *The Changing Experience of Employment: Restructuring and Recession*, London: Macmillan.
—— (1987) 'Redundancy and the labour market: the role of "readaptation benefits"', in R. Lee (ed.) *Redundancy, Layoffs and Plant Closures*, Beckenham: Croom Helm.
Fleming, D. K. (1967) 'Coastal steelworks in the Common Market countries', *Geographical Review* 57: 48–72.
Fleming, D. K. and Krumme, G. (1968) 'The Royal Hoesch Union: case analysis of adjustment patterns in the European steel industry', *Tijdschrift* 59: 177–99.
Gachelin, C. (1980) 'La mutation de la sidérurgie de la région du Nord', *Hommes et Terres du Nord* 1: 21–33.
Gauche-Cazalis, C. (1979) 'Sidérurgie: étatiser pour mieux casser', *Economie et Politique* 21: 28–33.
Grabitz, E. and Hanlon, G. (1984) 'A review of the steel quota cases: judicial endorsement of ECSC crisis management', *Common Market Law Review* 21: 163–222.
Green, D. (1983) 'Strategic management and the state: France', in K. Dyson and S. Wilks (eds) *Industrial Crisis: A Comparative Study of the State and Industry*, Oxford: Blackwell.
Gwynne, R. N. and Giles, B. D. (1980) 'The concept of Fos — an interim assessment', *Norsk Geografisk Tidsskrift* 34: 35–43.
Hall, P. G. (1981) 'Enterprise zones: British origins, American adaptation', *Built Environment* 7: 5–12.
Hartshorne, R. (1928) 'Location factors in the iron and steel industry', *Economic Geography* 4: 241–52.
Hogan, W. T. (1983) *World Steel in the 1980s: A Case of Survival*, Lexington, Mass: D.C. Heath.
Hudson, R. (1986) 'Producing an industrial wasteland: capital, labour and the state in north-east England', in R. Martin and B. Rowthorne (eds) *The Geography of De-Industrialisation*, London: Macmillan.
—— (1989) *Wrecking a Region: State Policies, Party Politics and Regional Change in North East England*, London: Pion.
Hudson, R. and Lewis, J. (eds) (1982) *Regional Planning in Europe*, London: Pion.
Hudson, R. and Sadler, D. (1983) 'Anatomy of a disaster: the closure of Consett steelworks', *Northern Economic Review* 6: 2–17.
—— (1984) *British Steel Builds the New Teesside? The Implications of BSC Policy for Cleveland*, Middlesbrough: Cleveland County Council.
—— (1985a) *The Development of Middlesbrough's Iron and Steel Industry 1841–1985*, Middlesbrough Locality Study WP2, Durham University.
—— (1985b) 'The past, present and future significance of British Steel's

policies for Cleveland', *Northern Economic Review* 11: 2–12.
—— (1986) 'Contesting works closures in Western Europe's old industrial regions: defending place or betraying class?', in A. J. Scott and M. Storper (eds) *Production, Work, Territory*, Boston, Mass: Allen & Unwin.
—— (1987a) 'Manufactured in the UK? Special steels, motor vehicles and the politics of industrial decline', *Capital and Class* 32: 55–82.
—— (1987b) *The Uncertain Future of Special Steels: Trends in the Sheffield, UK and European Special Steels Industries*, Sheffield City Council.
IISI (International Iron and Steel Institute) (1983) *Steel and the Automotive Sector*, Brussels.
—— (1985) *Indirect Trade in Steel*, Brussels.
Industry and Trade Committee (1983) *The British Steel Corporation's Prospects*, House of Commons Paper 212, Session 1982/83.
ISTC (Iron and Steel Trades Confederation) (1980) *New Deal for Steel*, London: Penshurst Press.
—— (1984) *Phoenix Two: The Threat to Engineering Steels*, London.
Jäger, S. (1981) 'Dortmund kann Lehrstück sein', in Revier-Redaktion (ed.) *Brennpunkt Stahlkrise*, Duisburg: Revier.
Jones, K. (1979) 'Forgetfulness of things past: Europe and the steel cartel', *World Economy* 2: 139–54.
—— (1985) 'Trade in steel: another turn in the protectionist spiral', *World Economy* 8: 393–408.
—— (1986) *Politics versus Economics in the World Steel Trade*, London: Allen & Unwin.
Junkerman, J. (1987) 'Blue-sky management: the Kawasaki story', in R. Peet (ed.) *International Capitalism and Industrial Restructuring*, London: Allen & Unwin.
Keating, M. and Hainsworth, P. (1986) *Decentralisation and Change in Contemporary France*, Aldershot: Gower.
Kourchid, O. (1987) 'Workers' struggles in steel in France and in the USA: autonomy and constraint at Longwy, Lorraine and at Youngstown, Ohio', in R. Lee (ed.) *Redundancy, Layoffs and Plant Closures*, Beckenham: Croom Helm.
Local Authority Associations (1986) *Local Authorities' Response to Decline of Major Traditional Industries*, London: LAA.
Manwaring, T. (1981) 'Labour productivity and the crisis at BSC: behind the rhetoric', *Capital and Class* 14: 61–97.
Markusen, A. (1985) *Profit cycles, oligopoly and regional development*, Cambridge, Mass: MIT.
—— (1986) 'Neither ore, nor coal, nor markets: a policy-oriented view of steel sites in the USA', *Regional Studies* 20: 449–62.
Martin, J. E. (1957) 'Location factors in the Lorraine iron and steel industry', *Transactions of the Institute of British Geographers* 23: 191–212.
Martin, R. and Rowthorne, B. (eds) (1986) *The Geography of De-Industrialisation*, London: Macmillan.

Maunders, A. (1987) *A Process of Struggle: The Campaign for Corby Steelmaking in 1979*, Aldershot: Gower.

Mény, Y. (1983) 'Permanence and change: the relations between central government and local authorities in France', *Government and Policy* 1: 17–28.

Morgan, K. and Sayer, A. (1985) 'A "modern" industry in a "mature" region: the remaking of management–labour relations', *International Journal of Urban and Regional Research* 9: 383–404.

Moro, D. (1984) *Crisis e ristruttorazione dell'industria siderurgica italiana*, Varese, Giuffre Editore.

Morris, L. D. (1987) 'Local social polarisation: a case-study of Hartlepool', *International Journal of Urban and Regional Research* 11: 331–50.

Mueller, H. and Van den Ven, H. (1982) 'Perils in the Brussels-Washington steel pact of 1982', *World Economy* 5: 259–78.

NAO (National Audit Office) (1986) *Department of the Environment, Scottish Office and Welsh Office: Enterprise Zones*, London: HMSO.

Nationalised Industries Committee (1977) *BSC and Technological Change*, House of Commons Paper 26 (three volumes), Session 1977/78.

NEDO (Plastics Processing EDC) (1985) *Replacement of Metals with Plastics*, London.

NEDO (1986a) *Steel: The World Market and the UK Steel Industry*, London.

—— (1986b) *Changing Working Patterns: How Companies Achieve Flexibility to Meet New Needs*, London.

Noiriel, G. (1980) *Vivre et lutter à Longwy*, Paris: Maspéro.

OECD (1985) *World Steel Trade Developments 1960–1983*, Paris: OECD.

Public Accounts Committee (1985) *Control and Monitoring of Investment by British Steel Corporation in Private Sector Companies — The Phoenix Operations*, House of Commons Paper 307, Session 1984/85, London.

Purcell, K., Wood, S., Watson, A., and Allen, S. (eds) (1986) *The Changing Experience of Employment: Restructuring and Recession*, London: Macmillan.

Région Nord-Pas de Calais (1984) 'A propos du Bassin Minier du Nord-Pas de Calais', *Note d'Information Economique* 73, Lille.

—— (1985) 'Le Plan Fabius', *Lettre de la Région N-PdC*, Lille.

Robinson, F. and Sadler, D. (1984) *Consett after the Closure*, O-P 19, Department of Geography, University of Durham.

—— (1985) 'Routine action, reproduction of social relations and the place market: Consett after the closure', *Society and Space* 3: 109–20.

Roger Tym and Partners (1984) *Monitoring Enterprise Zones: Year Three Report*, London.

Routledge, P. (1980) 'Why it was so worthwhile', *ISTC Banner*, July.

Sadler, D. (1984) 'Works closure at British Steel and the nature of the state', *Political Geography Quarterly* 3: 297–311.

—— (1986) 'Born in a steeltown: class relations and the decline of the European Community steel industry since 1974', unpublished Ph.D. thesis, University of Durham.

Schröter, L. (1982) '"Steelworks now!": the conflicting character of modernisation', paper presented to International Institute of Management Conference, Berlin.
Scott, A. J. and Storper, M. (1989) 'The geographical foundations and social regulation of flexible production systems', in J. Wolch and M. Dear (eds) *Territory and Social Reproduction*, London: Allen & Unwin.
Scottish Affairs Committee (1982) *The Steel Industry in Scotland*, House of Commons Paper 22, Session 1982/83.
—— (1985) *The Proposed Closure of BSC Gartcosh*, House of Commons Paper 154, Session 1985/86, London.
Sheffield City Council (1984a) *Steel in Crisis: Alternatives to Government Policy and the Decline in South Yorkshire's Steel Industry*, Sheffield.
—— (1984b) *An Employment and Environmental Plan for the Lower Don Valley*, Sheffield.
—— (1986) *Sheffield: Putting You in the Picture: Sheffield City Council, 1980–1986*, Sheffield.
—— (1987) *Sheffield: Working it Out: — An Outline Employment Plan for Sheffield*, Sheffield.
Shutt, J. (1984) 'Tory Enterprise Zones and the Labour Movement', *Capital and Class* 23: 19–44.
Sirs, B. (1977) 'Ten years of the TUC Steel Committee', *British Steel* 33: 12–15.
Storey, D. (1986) *Manufacturing Employment Change in County Durham 1965–1985*, Newcastle-upon-Tyne: CURDS.
Todd, G. (1984) *Creating New Jobs in Europe: How Local Initiatives Work*, London: Economist Intelligence Unit Special Report 165.
Trade and Industry Committee (1984) *The British Steel Corporations's Prospects*, House of Commons Paper 344, Session 1983/84.
Upham, M. (1980) 'British Steel: retrospect and prospect', *Industrial Relations Journal* 11: 5–21.
Walter, I. (1979) 'Protection of industries in trouble: the case of iron and steel', *World Economy* 2: 155–88.
—— (1983) 'Structural adjustment and trade policy in the international steel industry', in W. R. Cline (ed.) *Trade Policy in the 1980s*, Washington, DC: Institute of International Economics.
Ward, T. and Rowthorne, B. (1979) 'How to run a company and run down an economy: the effects of closing down steel-making in Corby', *Cambridge Journal of Economics* 3: 327–40.
Warren, K. (1967) 'The changing steel industry of the European Common Market', *Economic Geography* 43: 314–32.
Weston, P. G. (1985) 'Corby — the myth and the reality', unpublished MS, Northampton: NUPE.
Zukin, S. (1985) 'Markets and politics in France's declining regions', *Journal of Policy Analysis and Management* 5: 40–57.
Zum Brunnen, C. and Osleeb, J. P. (1986) *The Soviet Iron and Steel Industry*, Beckenham: Croom Helm.

Author index

Subject index